INTRODUCTION TO THE MECHANICS OF THE SOLAR SYSTEM

By the same author:

VON DEN GRENZEN DES WISSENS
(Basel 1953)

ZUM WELTBILD DER ASTRONOMIE
(With M. Schürer. Bern 1954 and 1957)

ENTWURF EINER METAPHYSIK
(Bern 1955)

CHRISTENTUM UND STAAT
(Bern 1957)

INTRODUCTION TO THE MECHANICS OF STELLAR SYSTEMS
(London 1957)

INTRODUCTION TO

THE MECHANICS OF THE SOLAR SYSTEM

RUDOLF KURTH

Department of Astronomy
University of Manchester

PERGAMON PRESS

NEW YORK · LONDON · PARIS · LOS ANGELES

1959

PERGAMON PRESS INC.
122 East 55th Street, New York 22, N.Y.
P.O. Box 47715, Los Angeles, California
PERGAMON PRESS LTD.
4 & 5 Fitzroy Square, London W.1.
PERGAMON PRESS S.A.R.L.
24 Rue des Écoles, Paris Ve

Library of Congress Card No. 59–12064

Printed in Northern Ireland at The Universities Press, Belfast

PREFACE

THIS book is perhaps somewhat unconventional. One of the principal reasons is that, like Socrates, the author dislikes long speeches—and long formulae; and the formulae of celestial mechanics can indeed be very long. No doubt these formulae are very useful or necessary for predicting the motions of the planets; I am not, however, interested in that here, but in understanding the fundamental principles and methods. Anyone, therefore, wishing to copy out ready-made recipes from this book will probably consult it in vain; but I hope that it will be possible to learn a methodical approach from it. My first concern has been with the needs of young students. I have had the following aims:

(i) Nearness to natural reality—as is seen, for example, in the occasional use of such names as Jupiter and Saturn.

(ii) Insight through simplicity: direct elementary methods and simply formulated results, expressed by short approximate formulae, as is both usually and fruitfully done in theoretical physics. The following requirements were made:

a. The results must be methodically derived.

b. They must be able to be interpreted easily.

c. They shall correctly describe the phenomena both qualitatively, and to the right order of magnitude quantitatively.

d. They must be capable of being improved to arbitrary accuracy, for example by iterative procedures or the inclusion of higher terms, by *the same* method by which they were derived. (Insight is obtained from the approximate formulae which are then improved for the needs of astronomical practice.)

(iii) Rigour: although proofs of convergence have been at most hinted at, I hope that at least a breath of the *esprit de géométrie* may be felt in this book.

(iv) Development of the basic concepts: for example the laws of motion and of gravitation were not to be presented as having fallen down ready-made from Heaven, but were to be derived from the observations.

(v) Approach by an axiomatic treatment: each experience or hypothesis should be exhausted before any new experiences or ideas are introduced. For didactical reasons, however, this maxim is not applied rigorously.

I should add to (ii) that I have not discussed Hamilton–Jacobi Theory. It seems to me, after detailed critical examination, that it is a vast detour from the way to the perturbation equations. The beauty of its general formulation (so simple in principle) is rather useless in face of the concrete problem of integrating a given system of ordinary differential equations. It is only in statistical problems—which are outside traditional celestial mechanics, and so beyond the scope of this book—that canonic variables offer a decided advantage, in conjunction with Liouville's Theorem.

So much for my intentions. Little previous knowledge is expected of the reader beyond the elements of analysis, analytical geometry and linear algebra. My references to literature are of a more random nature. Everything I quote I know from my own studies. The examples are intended to stimulate the reader to think for himself.

The greater part of this book was given as a lecture course in the University of Manchester. Professor Max Schürer and Mr. Jean-Paul Marchand, of Bern, have read critically a part of the manuscript. Dr. A. H. Batten, of Manchester, has translated it into English. To them all I offer my best thanks.

CHEADLE, Cheshire RUDOLF KURTH
8 February 1958.

CONTENTS

I. THE KINEMATICS OF A SINGLE PLANET

II. THE DYNAMICS OF A SINGLE PLANET

III. THE DYNAMICS OF THE PLANETARY SYSTEM

IV. THE PLANETS AND THE MOON AS RIGID BODIES

CHAPTER I

THE KINEMATICS OF A SINGLE PLANET

1. The periods of the planets

1. When, in 1600, Kepler applied himself to the examination of the orbit of Mars, he had at his disposal all the positional observations that Tycho Brahe had made over many years. His aim was an exact theoretical description of the planetary motions. Like all his predecessors, he assumed, as a matter of course, that the motions of the single planets are periodic, and that each planet always moves in the same closed orbital curve. These assumptions were compatible with observation, within the limits of experimental error at that time. We shall also take them as our basis in this chapter. Kepler had first to determine, from the observations, the period of Mars around the Sun. How that can be done will be shown in this first section.

2. First of all, however, we must make clear in what way the directions of a planet, relative to the Sun, may in fact be specified. For Kepler the matter was simple: he could, without further consideration, specify directions by means of the fixed stars. Within the limits of the accuracy of measurement in his time, the fixed stars could be regarded as forming rigid constellations. For us to-day it is different. We know that the fixed stars are moving relatively to each other and, for the nearer ones, we can even measure the relative lateral displacements. To lay down "fixed" directions (i.e. directions, that we postulate as fixed) we must therefore use stars whose angular motions relative to one another are too small for us to measure. So, the most distant stars, the globular clusters or even the extra-galactic star-systems can serve to fix the "fundamental directions." Arbitrary directions may then be described by the angles they make with the fundamental directions. It is obvious that three non-coplanar fundamental directions are sufficient to describe all directions. For example, we take a fundamental plane through two fundamental directions and define an azimuth angle in it; by means of an angle of latitude we express the angular distance

from the fundamental plane. The third direction serves, only qualitatively, to distinguish between the two sides of the fundamental plane (to give an orientation to it.) The azimuth and latitude angles together determine each direction uniquely.

Of course, it is possible that the part of the universe we observe is rotating with respect to some unobserved (and perhaps much larger) part, so that our apparently fixed directions may not be really fixed at all. But what do we mean by "really"? We can only establish the motions of the heavenly bodies relatively to each other, and that is the only sense in which the idea of motion is clear at all. We shall now relate all planetary motions to the fixed stars (in the sense discussed above). The question whether this system rotates relatively to some distant part of the universe is quite unimportant for the mechanics of the planetary system.

In astronomical theory and practice, directions have been established hitherto, for historical reasons and on grounds of practical convenience, in a different way; namely, with respect not only to the sphere of the fixed stars but also to the Earth. Because the fundamental ideas are shown more clearly, we prefer our own definition here.

3. We return to the problem of determining the period of a planet, let us say Mars, from the observations. We set out here the assumptions:

(i) The Earth and Mars move around the Sun in closed orbits which both lie in the same fixed plane through the Sun. (In reality, the orbital planes of the Earth and Mars are different, but then, instead of the actual motion of Mars, we can consider its projection on the Earth's orbital plane. Even the assumption that the orbital curve itself lies in a plane is unnecessary.)

(ii) The radius vector Sun–projection of planet and the radius vector Sun–Earth are one-valued functions of the direction.

(iii) The angular velocity of the projection of the planet relative to the Sun is a one-valued function of the direction relative to the Sun. In other words, the angular velocity is a periodic function $n(t)$ of the time t, whose period P coincides with the sidereal period, i.e. with the length of time between two consecutive passages of the planet through the same point of its orbit (relative to the Sun and the fixed axes in space).

Now at some time Mars will be in opposition to the Sun—i.e. the Sun, the Earth and Mars are situated, in that order, on a straight line. At the next moment this will no longer be true. However, oppositions will continually recur, let us say at the times $t_1, t_2, \dots$. These times were known to Kepler from observations over a long interval.

We now distinguish quantities concerning the Earth and Mars by the suffixes E and M respectively. Then observations at the time of opposition tell us:

$$0 < n_{\mathrm{M}}(t) < n_{\mathrm{E}}(t) \tag{1}$$

On this result we base our last assumption:

(iv) The relation (1) holds for all times t.

We deduce therefore that Mars is continually lagging behind the Earth (as seen from the Sun) which will overtake it at times $t_i (i = 0, 1, 2, \dots)$. The time interval $t_i \dots t_{i+1}$ we call a synodic period of Mars; its length $S_i = t_{i+1} - t_i$ is the duration of a synodic period. This is not constant (although only weakly variable), but a function of i.

From assumption (iv) it follows that

$$\int_{t_i}^{t_{i+1}} [n_{\mathrm{E}}(t) - n_{\mathrm{M}}(t)] \, \mathrm{d}t = 2\pi, \qquad i = 0, 1, 2, \dots$$

and, by addition of the first k of these equations,

$$\int_{t_0}^{t_k} [n_{\mathrm{E}}(t) - n_{\mathrm{M}}(t)] \, \mathrm{d}t = 2\pi k, \qquad k = 0, 1, 2, \dots \tag{2}$$

We write now

$$t_k - t_0 = \begin{cases} P_{\mathrm{E}} \, . \, (N_{\mathrm{E}} + r_{\mathrm{E}}), \\ P_{\mathrm{M}} \, . \, (N_{\mathrm{M}} + r_{\mathrm{M}}) \end{cases} \tag{3}$$

where N_{E} is the greatest whole number $\leqslant (t_k - t_0)/P_{\mathrm{E}}$ and r_{E} is the corresponding remainder; N_{M} and r_{M} are similarly defined. All four numbers are uniquely determined by $t_k - t_0$, P_{E} and P_{M}. Moreover,

$$0 \leqslant r_{\mathrm{E}} < 1, \qquad 0 \leqslant r_{\mathrm{M}} < 1. \tag{4}$$

The quantities N_{E} and r_{E} are known empirically; N_{M} and r_{M} are the quantities we wish to determine. Obviously, $(N_{\mathrm{E}} + r_{\mathrm{E}})$ is the number of terrestrial years which have elapsed in the relevant k synodic periods, while $(N_{\mathrm{M}} + r_{\mathrm{M}})$ is the corresponding number of Martian years.

The integral

$$\int_{t_0}^{t_k} [n_{\mathrm{E}}(t) - n_{\mathrm{M}}(t)] \, \mathrm{d}t$$

on the left-hand side of (3) may now be calculated by the following method. We evaluate $\int n_{\mathrm{E}}\,\mathrm{d}t$ and $\int n_{\mathrm{M}}\,\mathrm{d}t$ separately. Then, to determine the first of these integrals, we separate the interval of integration $t_0 \ldots t_k$ into the N_{E} partial intervals $t_0 \ldots t_0 + P_{\mathrm{E}}$, $t_0 + P_{\mathrm{E}} \ldots t_0 + 2P_{\mathrm{E}}, \ldots,$ $t_0 + (N_{\mathrm{E}} - 1)P_{\mathrm{E}} \ldots t_0 + N_{\mathrm{E}}P_{\mathrm{E}}$ and into the "remainder interval" $t_0 + N_{\mathrm{E}}P_{\mathrm{E}} \ldots t_0 + (N_{\mathrm{E}} + r_{\mathrm{E}})P_{\mathrm{E}}$. Each of the first N_{E} intervals makes a contribution of 2π to the total integral (see assumption (iii)), and the "remainder interval" contributes $2\pi\alpha_{\mathrm{E}}$ where

$$0 \leqslant \alpha_{\mathrm{E}} < 1. \tag{5}$$

So we have

$$\int_{t_0}^{t_k} n_{\mathrm{E}}(t)\,\mathrm{d}t = 2\pi(N_{\mathrm{E}} + \alpha_{\mathrm{E}})$$

and, similarly,

$$\int_{t_0}^{t_k} n_{\mathrm{M}}(t)\,\mathrm{d}t = 2\pi(N_{\mathrm{M}} + \alpha_{\mathrm{M}}),$$

$$0 \leqslant \alpha_{\mathrm{M}} < 1. \tag{6}$$

Setting this in equation (2) we obtain

$$N_{\mathrm{E}} - N_{\mathrm{M}} - k = \alpha_{\mathrm{M}} - \alpha_{\mathrm{E}}.$$

Since the left-hand side of this equation is a whole number so also must be the right-hand side. By (5) and (6), however,

$$-1 < \alpha_{\mathrm{M}} - \alpha_{\mathrm{E}} < +1.$$

Therefore

$$\alpha_{\mathrm{M}} - \alpha_{\mathrm{E}} = 0$$

and

$$N_{\mathrm{E}} - N_{\mathrm{M}} = k, \tag{7}$$

hence N_{M}.

If we knew r_{M} as well we could determine the required period P_{M} from equation (3). The exact evaluation of r_{M}, however, is not possible without further assumptions. Since, however, r_{M} is limited by (4) while N_{M}, on the other hand, increases indefinitely with time, an asymptotic estimate may be given for P_{M}, namely:

$$P_{\mathrm{M}} = \frac{t_k - t_0}{N_{\mathrm{M}} + r_{\mathrm{M}}} = \frac{t_k - t_0}{N_{\mathrm{M}} + \frac{1}{2}} \cdot \frac{1 + \dfrac{1}{2N_{\mathrm{M}}}}{1 + r_{\mathrm{M}}/N_{\mathrm{M}}}$$
$$= \frac{t_k - t_0}{N_{\mathrm{M}} + \frac{1}{2}}\left[1 + (\tfrac{1}{2} - r_{\mathrm{M}})\frac{1}{N_{\mathrm{M}}} + O\left(\frac{1}{N_{\mathrm{M}}^2}\right)\right]. \tag{8}$$

This means that if P_{M} is calculated with a mean value $r = \frac{1}{2}$ then the upper limit for the relative error is approximately equal to $1/2N_{\mathrm{M}}$.

The relative error of this result may be considerable. For example for $N_{\mathrm{M}} = 10$ the period of Mars may be determined only with an accuracy of $\frac{1}{20} = 5\%$. Under some circumstances, however, it is possible to narrow the margin of error much further. How this may be done will be shown in the following subsection.

4. At this point we introduce the mean synodic period defined by the equation

$$\bar{S}_k = \frac{S_0 + S_1 + \ldots + S_{k-1}}{k} = \frac{t_k - t_0}{k}, \; k = 1, 2, \ldots \tag{1}$$

From equation § 1.3(8) we have:

$$P_{\mathrm{M}} = \frac{k\bar{S}_k}{N_{\mathrm{M}} + r_{\mathrm{M}}}$$

or, by § 1.3(7)

$$P_{\mathrm{M}} = \frac{k\bar{S}_k}{N_{\mathrm{E}} - k + r_{\mathrm{M}}}.$$

If we take reciprocals we have

$$\frac{1}{P_{\mathrm{M}}} = \frac{N_{\mathrm{E}} + r_{\mathrm{M}}}{k\bar{S}_k} - \frac{1}{\bar{S}_k}$$

$$= \frac{1 + \dfrac{r_{\mathrm{M}}}{N_{\mathrm{E}}}}{\dfrac{t_k - t_0}{N_{\mathrm{E}}}} - \frac{1}{\bar{S}_k}$$

$$= \frac{1}{P_{\mathrm{E}}} \frac{1 + \dfrac{r_{\mathrm{M}}}{N_{\mathrm{E}}}}{1 + r_{\mathrm{E}}/N_{\mathrm{E}}} - \frac{1}{\bar{S}_k}$$

by § 1.3(3). This may be written

$$\frac{1}{P_{\mathrm{M}}} = \frac{1}{P_{\mathrm{E}}} - \frac{1}{\bar{S}_k} + \varepsilon_k, \tag{2}$$

where ε_k is given by

$$\frac{1}{P_E} + \varepsilon_k = \frac{1}{P_E} \cdot \frac{1 + \dfrac{r_M}{N_E}}{1 + \dfrac{r_E}{N_E}} = \frac{1}{P_E}\left[1 + O\left(\frac{1}{N_E}\right)\right];$$

hence

$$\varepsilon_k = O\left(\frac{1}{N_E}\right). \tag{3}$$

From this we deduce that as $k \to \infty$ there exists a limiting value

$$\operatorname*{Lim}_{k \to \infty} \bar{S}_k.$$

We denote this limit by $\bar{S}$ and call it "the" synodic period of the planet. For this, the relation

$$\frac{1}{P_M} = \frac{1}{P_E} - \frac{1}{\bar{S}} \tag{4}$$

holds.

Now, if the empirical values of $\bar{S}_1, \bar{S}_2, \bar{S}_3, \ldots, \bar{S}_k$ oscillate within only a narrow range, then, assuming k is a large number, it may be supposed that the limiting value $\bar{S}$ also lies in this range. Under some circumstances the last formula may be used to obtain a definite value of the sidereal period by exact calculation. For instance, consider the case of constant angular velocities $n_E(t)$ and $n_M(t)$.

If the assumed value of $\bar{S}_M$ is in error by $\delta\bar{S}_M$ then, by (4), the corresponding error δP_M is, to the first approximation,

$$\frac{\delta P_M}{P_M^2} = -\frac{\delta\bar{S}_M}{\bar{S}_M^2}$$

or

$$\frac{\delta P_M}{P_M} = -\frac{P_M}{\bar{S}_M} \cdot \frac{\delta\bar{S}_M}{\bar{S}_M} \tag{5}$$

or by (4)

$$\frac{\delta P_M}{P_M} = -(P_M - 1)\frac{\delta\bar{S}_M}{\bar{S}_M},$$

if the terrestrial year, P_E, is taken as unit of time. From this equation we see that, for planets whose sidereal periods are larger than two, the relative error in the sidereal period is greater than that in the synodic period.

5. For those planets that move with a greater angular velocity than the Earth it is obviously necessary to substitute $-k$ for k and $-S$ for S throughout the formulae of the two preceding sections. Only Mercury and Venus are faster (in this sense) than the Earth. For these planets, however, neither the original nor the modified formula can be applied, since they are never in opposition to the Sun. In particular, the definition of the synodic period is no longer applicable. Conjunctions (i.e. linear configurations of the Earth, Sun and planet or planet-projection with the Earth at one end certainly occur, and could be used for a new definition if it were not that the light of the Sun hampers their being observed. So we must define the synodic period in some other way, for example by the interval between two successive times when the planet (or its projection) is at its greatest western or (if preferred) eastern angular distance from the Sun. To determine the sidereal period from this synodic period we suppose first that the Earth and (let us say) Venus (or its projection) move in co-planar and concentric circular orbits around the Sun. Then the angular velocity of Venus, around the Sun, relative to the radius vector Sun–Earth is $n_{\mathrm{V}} - n_{\mathrm{E}}$. The time between two similar configurations, that is the synodic period, is then given by

$$(n_{\mathrm{V}} - n_{\mathrm{E}})S = 2\pi$$

that is

$$n_{\mathrm{V}} - n_{\mathrm{E}} = \frac{2\pi}{S}$$

or

$$\frac{1}{P_{\mathrm{V}}} - \frac{1}{P_{\mathrm{E}}} = \frac{1}{S}.$$

It is geometrically evident, and can easily be shown analytically, that for motions differing only slightly from those in concentric circular orbits, the relation

$$\frac{1}{P_{\mathrm{V}}} - \frac{1}{P_{\mathrm{E}}} = \frac{1}{\bar{S}} \tag{1}$$

holds, where $\bar{S}$ is, again, the limiting value of the mean synodic period.

6. We now present a tabular summary of the synodic and sidereal periods of the planets. The unit is a terrestrial year.

A few comments about the discovery of the planets are appended: Mercury, Venus, Mars, Jupiter and Saturn have been known from antiquity. Uranus was discovered observationally, by William Herschel, in 1781. The existence of Neptune was

TABLE 1

Planet	Synodic period	Sidereal period	Year of discovery
Mercury	0·31726	0·24084	—
Venus	1·59866	0·61519	—
Earth	—	1·00000	—
Mars	2·13532	1·88082	—
Minor planets	ca. 2–1·1	ca. 2–9	since 1801
Jupiter	1·09206	11·86178	—
Saturn	1·03514	29·45660	—
Uranus	1·01206	84·015	1781
Neptune	1·00612	164·77	1846
Pluto	1·00406	248·3	1930

inferred from perturbations in the motion of Uranus. In 1846, the motion of the supposed planet was computed, from these perturbations, by Adams in Cambridge and by Leverrier in Paris. The planet itself was observed, in the same year, by Galle in Berlin, with the aid of Leverrier's computations. Again, perturbations of Uranus convinced Lowell of the existence of Pluto and he computed its orbit. In 1930 the planet was found near the predicted position. The first of the Minor Planets (Ceres) was discovered observationally, in 1801, by Piazzi in Palermo. To-day several thousand are known.

Taking into account the dates of discovery of both the last two planets and the length of their sidereal periods we see that, as yet, they have been observed for only a fraction of their whole sidereal periods. It may be, therefore, that the present observed values of $\bar{S}_k$ still differ quite considerably from $\bar{S}$. By § 1.4(5) the relative error of the sidereal period will be quite large.

2. Kepler's determination of the orbits of the Earth and Mars: his first law

1. In this section we shall discuss Kepler's method of determining orbits—albeit with some modifications. Since we are

interested only in the principle we shall consider only the geometrical core of the method without discussing any of the details of the calculations. Further, for simplicity we shall again confine ourselves to a plane construction in which we consider the projection of the motion of Mars on to the Earth's orbital plane. The generalization to three dimensions offers no difficulties of principle. Finally, we modify Kepler's ideas yet again in that we do not follow his roundabout way, but travel directly to his goal.

The problem falls, then, into two parts: first, the orbit of the Earth must be constructed and from this the orbit of Mars. For this purpose we can assume that the direction of the Earth as seen from the Sun and the direction of Mars as seen from the Earth are empirically known functions of the time throughout a long interval. As basic hypotheses we take assumptions (i)–(iv) of § 1.3 and, in particular the assumption that both orbits are closed curves. The sidereal period of Mars may be determined from the empirical data by the method of the preceding section and, therefore, it will also be assumed as empirically known.

2. To construct the orbit of the Earth we start from an opposition of Mars at time t_0. If P is the sidereal period of Mars, then at the times t_0, $t_0 + P$, $t_0 + 2P$ ... Mars is at the same point in space. This is not so for the Earth, however. The corresponding positions for the Earth may be deduced from two data: the angle Earth–Sun–Mars, and the angle Sun–Earth–Mars, both obtained from observations of the Sun and of Mars. (See Fig. 1.) The distance Sun–Mars is taken as the preliminary unit of length. After the orbit of the Earth has been constructed its measure may be used to define the unit of length. Because all measurements are angular measurements we cannot obtain absolute distances, but only their ratios.

We insert, here, a note of a purely mathematical character. If the ratio P/P_0 of the periods of Mars and the Earth is irrational then we have infinitely many different positions for the Earth, and these form a dense subset of its orbit so that any arbitrarily small element of arc in the orbit contains at least one of the constructed positions. From considerations of continuity, therefore, we see that the whole orbit is uniquely determined by Kepler's construction. On the other hand, if the ratio is rational, the construction gives only a finite number of positions for the

Earth. The orbit is then no longer uniquely determined since we may interpolate between neighbouring positions as we please. Astronomically, this distinction naturally does not exist: P_0 and P are empirically determined numbers and as such are inexact. To speak of them or their ratio as rational or irrational

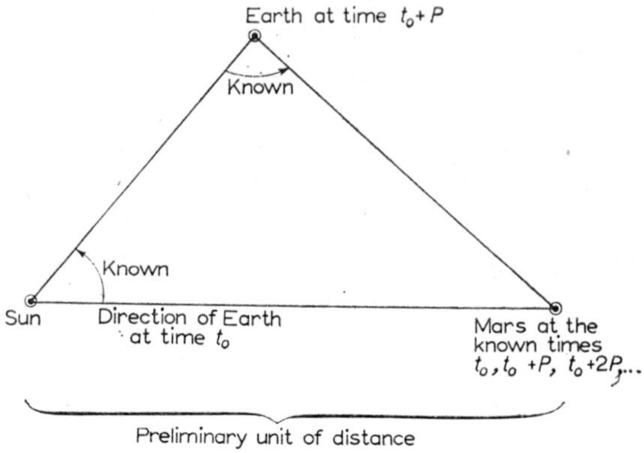

Fig. 1

is, therefore, meaningless. In fact we always obtain only a finite number of positions more or less exactly and find the rest by interpolation.

The result of Kepler's construction is that the orbit of the Earth is an ellipse (although nearly circular) with the Sun in one focus.

3. Now we turn to the construction of the orbit of Mars. For this we start from any fixed time t. Again at the times t, $t + P$, $t + 2P$ ... Mars occupies the same point in space. The positions of the Earth, E_0, E_1, E_2, ... at these instants are known, as also the directions in which Mars will be seen from them. Draw, now, straight lines through E_0 and E_1 corresponding to the directions of Mars—they will intersect in the position of Mars at time t. In this way the position of Mars at any time t may be found. (Only if P is a whole number does the method fail.)

Kepler found by a similar method that the orbit of Mars is an ellipse with the Sun in one focus. (Of course the eccentricity is small, but still markedly bigger than that of the orbit of the Earth.) From these two determinations of the orbits of the Earth and of Mars Kepler inferred, by induction, his first law:

all planetary orbits are ellipses with the Sun in one focus. Later investigations have confirmed the induction.

4. We now present in tabular form certain data for the planetary orbits. In particular we give the semi-major axes a, the numerical eccentricities e and ι, the inclinations of the various orbital planes to the plane of the orbit of the Earth.

TABLE 2

Planet	Semi-major axis a	Numerical eccentricity e	Ellipticity	Orbital inclination ι
Mercury	0·387	$2{\cdot}1 \times 10^{-1}$	$2{\cdot}2 \times 10^{-2}$	7°
Venus	0·723	$6{\cdot}8 \times 10^{-3}$	$2{\cdot}3 \times 10^{-5}$	3°
Earth	1	$1{\cdot}7 \times 10^{-2}$	$1{\cdot}4 \times 10^{-4}$	—
Mars	1·52	$9{\cdot}3 \times 10^{-2}$	$4{\cdot}3 \times 10^{-3}$	2°
Minor planets	*	up to 0·65	up to 0·21	up to 43°
Jupiter	5·20	$4{\cdot}8 \times 10^{-2}$	$1{\cdot}1 \times 10^{-3}$	1°
Saturn	9·54	$5{\cdot}6 \times 10^{-2}$	$1{\cdot}6 \times 10^{-3}$	2°
Uranus	19·2	$4{\cdot}7 \times 10^{-2}$	$1{\cdot}1 \times 10^{-3}$	1°
Neptune	30·1	$8{\cdot}6 \times 10^{-3}$	$3{\cdot}7 \times 10^{-5}$	2°
Pluto	39·5	$2{\cdot}5 \times 10^{-1}$	$3{\cdot}1 \times 10^{-2}$	17°

* Almost all between Mars and Jupiter.

Naturally, these modern values were not obtained by Kepler's method which requires a very large number of observations. Nonetheless, they could be found in this way, in principle, at least for the five planets of antiquity.

The "ellipticity" is defined as the quantity

$$\frac{a-b}{a},$$

where

$$b = \sqrt{(a^2 - (ae)^2)} = a\sqrt{(1-e^2)}$$

and denotes the semi-minor axis of the ellipse. The ellipticity is the relative difference between a and b, and so provides a more intuitive representation of the shape of the orbit than does the eccentricity itself. The ellipticity has been calculated, approximately, by the formula

$$\frac{a-b}{a} = 1 - \sqrt{(1-e^2)} = \tfrac{1}{2}e^2[1 + O(e^2)],$$

if $e \to 0$.

From the table we see:

(a) the orbital inclinations of all the planets, except those of Pluto and several minor planets, are very small so that the orbital planes nearly coincide.

(b) The eccentricities are so small that the orbits are nearly circular. Exceptions to this rule are, again, the Minor Planets, and, to a lesser extent, Pluto and Mercury. The eccentric position of the Sun, described by the value of e is, admittedly, considerable in some cases.

We add here two points which may not be seen from the table:

(c) all revolutions are described in the same sense, and

(d) the semi-major axes a may be approximately represented by the following formula:

$$a = 0{\cdot}3 \times 2^n + 0{\cdot}4.$$

In this formula n denotes the number of the planet if Venus is numbered as zero. (This is Titius–Bode's rule, discovered in 1766, i.e. before the discovery of Uranus, Neptune, Pluto and the Minor Planets.) In the following table we compare the empirical

TABLE 3

Planet	n	$0{\cdot}3 \times 2^n + 0{\cdot}4$	a (empirical value)
Mercury	$-\infty$	0·4	0·39
Venus	0	0·7	0·72
Earth	1	1·0	1
Mars	2	1·6	1·52
Minor planets	3	2·8	
Jupiter	4	5·2	5·20
Saturn	5	10·0	9·54
Uranus	6	19·6	19·2
Neptune	—	—	30·1
Pluto	7	38·8	39·5

values with those of the formula. The exceptions are Mercury with the number $-\infty$, and Neptune which does not fit in with any number at all. The Minor Planets, on the other hand, taken all together, are near to the correct place.

For the "theoretical" discoveries of Neptune and Pluto, the semi-axes a of the unknown planets were estimated by means of this rule. This seemed justifiable for Neptune, especially since the newly discovered Uranus obeyed the rule; but it is just in the case of Neptune that the rule failed.

Some authors place so much weight on these deviations that they deny any real significance to the Titius–Bode rule; with its three empirical numbers (0·3, 2, and 0·4) it is, in their opinion, nothing but a three-parameter interpolation formula. These reservations seem somewhat exaggerated, however, even if we cannot give it a physical interpretation. The actual form of the rule is not essential, of course, but it represents a law of the increase of semi-axes. From Mars outwards these are approximately doubled as we go from planet to planet.

3. Kepler's second and third laws

1. After Kepler had determined the orbit of Mars, he investigated the motion of the planet in its orbit. He evaluated the orbital velocity and found that it obeyed the following law: the product of the instantaneous velocity and the distance from the Sun of the corresponding tangent to the orbit, is constant. This can also be expressed: the radius vector from the Sun to the planet sweeps out equal areas in equal times. Or, yet again: *the areal velocity of the planet* (that is the derivative of the area swept out with respect to time) *is constant.* The inductive generalization made from this to all the planets is known as Kepler's second law.

That the three formulations are actually equivalent appears intuitively obvious from a consideration of two positions of Mars so close together that the arc joining them may be regarded, to a sufficient approximation, as the tangent at the first position. The analytical proof may be derived similarly.

2. Kepler had realized, quite early, that a relation must exist between the period, P, and the semi-major axis, a, of the various planets. The periods increase monotonically with the semi-axes. After some failures he found the expected relationship. It is his third law, which may be written: *the squares of the periods of two planets are in the same ratio as the cubes of the semi-major axes of their orbits.* In symbols we have

$$\frac{P^2}{P_1^2} = \frac{a^3}{a_1^3}.$$

The suffix one denotes a comparison planet. If we choose this to be the Earth, and its orbital semi-axis and period as units, the law is simply

$$P^2 = a^3.$$

We shall not give a numerical demonstration of this law by means of another table, since the generally accepted values of the semi-major axes (which we have quoted) have all been directly calculated from the sidereal periods by means of Kepler's third law itself. So these values cannot be used as a proof of the law. Data must be used that were obtained by such a method as Kepler's.

3. If the distance, in kilometres, between any two members of the solar system is known at some instant, then all its "absolute" measurements are known. Such an absolute dimension that may be obtained, for example, is the distance of any planet near enough to the Earth for trigonometrical measurements to be made (e.g. Mars, Venus or Eros—a Minor Planet which comes inside the orbit of Mars and very near to the Earth). For the base of the triangle a sufficiently large distance, of known length, is taken on the Earth. Naturally the relative orbits of the Earth and the planet must be well known. The calculation of orbits on the basis of Kepler's three laws will be dealt with in the next section.

In this way we obtain, for the semi-major axis a_E of the Earth's orbit, the value

$$a_E = 1{\cdot}5 \times 10^8 \text{ km}.$$

This, therefore, is the astronomical unit of length. The mean orbital velocity may be calculated from this and from the length of the sidereal year:

$$P_E = 3{\cdot}16 \times 10^7 \text{ sec}$$

so, approximately,

$$\frac{2\pi a_E}{P_E} \approx 29{\cdot}8 \text{ km sec}^{-1},$$

if, for the sake of simplicity, we take the nearly circular orbit of the Earth to be exactly circular.

In the astronomical system of units ($a_E = 1$, $P_E = 1$), which we prefer here, the velocity 2π corresponds, therefore, roughly to the value 30 km sec^{-1}. The astronomical unit of velocity is, therefore, 4·74 km sec^{-1}.

Conversely the semi-major axis may be estimated from measures of the Earth's orbital velocity. Observations may be made of the annual oscillation of the radial velocities of fixed stars which lie in the plane of the Earth's orbit. Apart from double stars and other exceptions, they must all show the same range. The Earth's orbital velocity is half of this. From this and the period an approximate value of the semi-major axis may be obtained.

4. We now bring Kepler's three laws together. They form the foundation for all that follows.

I. The orbits of the planets are ellipses with the Sun in one focus.

II. The areal velocity of each planet is constant with respect to time.

III. The square of the period of a planet is equal to the cube of the semi-major axis of its orbit.

4. The elements of a planetary orbit

1. We ask now what data are needed to specify uniquely the orbital ellipse of a planet in space.

To answer this question we must first introduce a suitable coordinate system. The Sun is taken as the origin of this system. Then we choose a mean plane of the Earth's orbit to be the (x, y)-plane. (We say "a mean plane" since it will later become clear that Kepler's laws hold only approximately. In particular, the orbit of a planet only approximates to an ellipse—or, indeed, to any plane curve. We define the "mean plane" of the Earth's orbit as the plane through the Sun and any two positions of the Earth. In any case the oscillations of the Earth's orbital plane are very small; they are a few seconds of arc in magnitude. We shall not consider such refinements until Chapter III. The direction of the x-axis is arbitrary.

2. We return now to the specification of a planet's orbit. If the orbital plane of the planet does not coincide with that of the Earth, then the planes intersect in a straight line called the "nodal line". (They cannot be parallel, since both planes have a common point—namely the centre of the Sun.) The points of intersection of the nodal line with the orbit are called nodes. The node at which the planet crosses from the south to the north side of the Earth's orbital plane is called the ascending node.

The other is the descending node The angle between the segment of the nodal line which includes the ascending node and the x-axis is the longitude of the node. It is denoted by Ω. In general, the "longitude" of a planet's position means the angle between the projection of the radius vector on the (x, y)-plane and the x-axis. Longitudes are measured in the direction of the Earth's motion.

We have already introduced the orbital inclination ι, (§2.4). It is the angle between the orbital planes of the planet and of the Earth. It is measured from 0 to 180° so that each pair of values (Ω, ι) corresponds to only one plane. (If ι were only measured to 90°, there would be, in general, two planes for each pair of values.)

We have now to characterize the orientation of the orbital ellipse in its plane. To do so we use ω, the angle between the direction of the ascending node and that of perihelion, the point of the orbit nearest to the Sun. (The point furthest from the Sun is called aphelion, both points are called apses and the line joining them is called the apsidal line.) The angle ω is measured from the ascending node, from 0 to 360° in the direction of motion.

The orientation of the orbital ellipse is determined by Ω, ι, ω. Its shape is uniquely described by the numerical eccentricity e, and its size by the semi-axis a. The orbital curve of the planet is, therefore, uniquely determined by the five "elements" Ω, ι, ω, e, a.

3. The coordinate system introduced in § 4.1 is useful if it is wished to consider the motion of several planets (of which the Earth may be one.) If we investigate only one planet, however, we use a new coordinate system in which the orbital plane of that particular planet is taken as a coordinate plane. We denote the old coordinates by x, y, z, and the new ones by ξ, η, ζ. The origin is, again, the Sun. We choose the direction of the positive ξ-axis to be that of the ascending node. Then the η-axis lies in the orbital plane and the positive ζ-axis makes the angle ι with the positive Z-axis. Thus we specify the (ξ, η, ζ) system uniquely.

We now give the transformation formulae between the two systems. We go from one to another as follows:

(a) From the (x, y, z)-system we obtain by a rotation about the z-axis through the angle Ω, an (x', y', z')-system. The x' axis is, therefore, the nodal line. The z'- and z-axes coincide.

(b) From the (x', y', z')-system we obtain the (ξ, η, ζ)-system by a rotation about the x'-axis through the angle ι. The nodal line is now the ξ-axis and the η-axis is in the orbital plane.

The resulting transformation reads:

$$\left.\begin{aligned} \xi &= x \cos \Omega + y \sin \Omega, \\ \eta &= -x \sin \Omega \cos \iota + y \cos \Omega \cos \iota + z \sin \iota, \\ \zeta &= x \sin \Omega \sin \iota - y \cos \Omega \sin \iota + z \cos \iota. \end{aligned}\right\} \tag{1}$$

The inverse transformation is given by

$$\left.\begin{aligned} x &= \xi \cos \Omega - \eta \sin \Omega \cos \iota + \zeta \sin \Omega \sin \iota, \\ y &= \xi \sin \Omega + \eta \cos \Omega \cos \iota - \zeta \cos \Omega \sin \iota, \\ z &= \eta \sin \iota + \zeta \cos \iota. \end{aligned}\right\} \tag{2}$$

Naturally, in the planet's orbit $\zeta = 0$.

4. We saw in § 4.2 that the elements

$$\Omega, \iota, \omega, e, a$$

specify the orbital curve of the planet in space. We now have to consider what constants are needed to determine the position of the planet in the orbit. At first sight it would seem that two are necessary, for example the time T of the planet's passage through perihelion and the constant c, the areal velocity of the planet.

The time of perihelion T and the areal velocity c determine uniquely the position of the planet. For at the time t the area swept out by the radius vector since perihelion is, by Kepler's second law, equal to $c(t - T)$. Obviously the time determines this area which in turn determines a point on the curve.

In fact, one of these constants suffices to specify uniquely the position of the planet, namely the time of perihelion T. The areal velocity c may be obtained from the geometrical elements by means of Kepler's laws in the following fashion.

During a complete revolution the radius vector sweeps out the whole area of the ellipse. Therefore

$$c = \frac{\text{area contained in the ellipse}}{\text{period}}.$$

We again use

$$b = a\sqrt{(1 - e^2)}$$

to denote the semi-minor axis of the orbital ellipse, so that the area contained in this ellipse is πab. Therefore

$$c = \frac{\pi ab}{P} = \frac{\pi a^2}{P} \sqrt{(1 - e^2)} \tag{1}$$

while, by Kepler's third law

$$P = a^{3/2}$$

so

$$c = \pi \sqrt{(1 - e^2)} \sqrt{a}. \tag{2}$$

We set forth, then, this principle: the six elements

$$\begin{cases} \Omega & \text{longitude of the node} \\ \iota & \text{orbital inclination} \\ \omega & \text{direction of perihelion} \\ e & \text{eccentricity} \\ a & \text{semi-major axis} \\ T & \text{time of perihelion} \end{cases}$$

uniquely determine the position of a planet at any time t.

5. The computation of the motion of a planet

1. We now pose a further problem: the elements of the orbit of a planet being given, to calculate its position at any instant t. To solve this, it is sufficient to evaluate the coordinates ξ, η in the orbital plane. We have already shown in § 4.4 how the coordinates x, y, z may be found from these. This coordinate transformation eliminates the elements Ω and ι so that, here, we have to deal only with the elements ω, e, a, T.

It is convenient to introduce polar coordinates, r, w in the orbital plane, whose origin is in the Sun. The "true anomaly", w, is measured in the same sense as the motion of the planet, and from the direction of perihelion. The angle between the radius vector (r, w) and the ξ-axis is $(\omega + w)$. Therefore the relation between the ξ, η Cartesian coordinates and the polar coordinates is

$$\left.\begin{aligned} \xi &= r \cos (\omega + w), \\ \eta &= r \sin (\omega + w). \end{aligned}\right\} \tag{1}$$

We now express Kepler's first and second laws symbolically. For the first law we obtain, from analytical geometry

$$r = a\,\frac{1 - e^2}{1 + e\cos w} \tag{2}$$

and for the second law

$$\tfrac{1}{2}r^2\,\frac{dw}{dt} = c. \tag{3}$$

Now, in equation (3) we substitute for r from equation (2) and also write (from § 4.4)

$$c = \pi\sqrt{a}\sqrt{(1 - e^2)}. \tag{4}$$

Thus we obtain

$$\tfrac{1}{2}a^2\left(\frac{1 - e^2}{1 + e\cos w}\right)^2\frac{dw}{dt} = \pi\sqrt{a}\sqrt{(1 - e^2)} \tag{5}$$

or

$$(1 - e^2)^{3/2}\int_0^w \frac{dw}{(1 + e\cos w)^2} = \int_0^t 2\pi\,\frac{dt}{a^{3/2}}\,. \tag{6}$$

There are now two problems to solve. First, the integral on the left-hand side of (6) must be evaluated to give t as a function $t(w)$ of w. Secondly, this equation must be solved for w to give w as a function $w(t)$ of t.

As far as the integration is concerned, we may develop the integrand in a (uniformly convergent) power series in e in each closed interval in the range $0 \leqslant e < 1$. The powers of $\cos w$ may be expressed in terms of $\cos w$, $\cos 2w$, $\cos 3w$ etc., and the series integrated term by term. We explain the process for the first term.

It follows from

$$\frac{1}{(1 + e\cos w)^2} = 1 - 2e\cos w + O(e^2)$$

that

$$\int_0^w \frac{dw}{(1 + e\cos w)^2} = w - 2e\sin w + O(e^2)\,.$$

(If it is required, an upper bound of the remainder term may be calculated exactly by Taylor's formula.) We now write

$$2\pi\,\frac{t - T}{a^{3/2}} = 2\pi\,\frac{t - T}{P} \equiv M \tag{7}$$

where M is called the mean anomaly. It is the angle that the direction of the planet would make with that of perihelion, were the planet to move with its mean constant angular velocity (with the "mean motion")

$$n = \frac{2\pi}{P} \tag{8}$$

around the Sun. Thus the perihelion direction corresponds to zero mean anomaly.

If we substitute all these expressions into the relation (6) between the true and mean anomalies, we obtain

$$w - 2e \sin w + O(e^2) = M. \tag{9}$$

Solving for w gives

$$\begin{aligned} w &= M + 2e \sin w + O(e^2) \\ &= M + 2e \sin M + O(e^2). \end{aligned} \tag{10}$$

The next higher term may be found similarly.

The integration

$$\int_0^w \frac{\mathrm{d}w}{(1 + e \cos w)^2} \tag{11}$$

may also be evaluated by means of elementary closed expressions, by substituting a new integration variable, $\tan \frac{1}{2}w$. This gives a rational function for the new integrand. This calculation, however, is rather tedious and, like Kepler, we prefer a geometrical treatment.

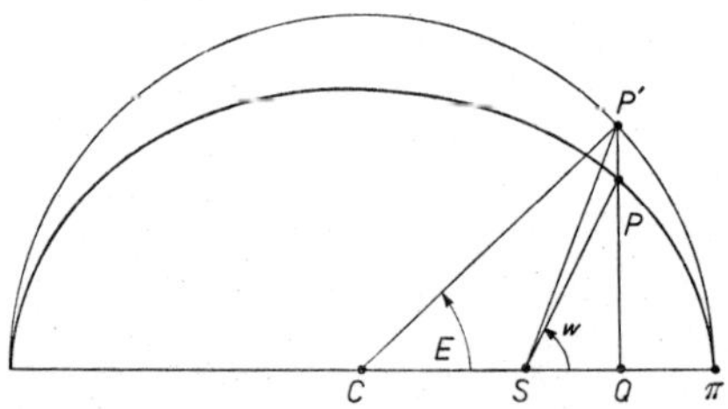

Fig. 2

2. In the orbital plane of the planet which we are considering, let us construct a circle of radius a whose centre is the centre C of the orbital ellipse. Let P be the position of the planet at the time t. We draw, through P, a line parallel to the minor axis of

the ellipse, which intersects the major axis in Q and the circle in P'. It is convenient to calculate, first, the so-called "eccentric anomaly" E, (the angle $P'CQ$). Obviously E and w determine each other uniquely and each may be calculated from the other.

Now, on the one hand, by Kepler's second law the area of the segment $PS\Pi$ is

$$\begin{aligned}(PS\Pi) &= c(t - T) \\ &= \frac{\pi ab}{P}(t - T) \qquad \text{(by § 4.4(1))} \\ &= \tfrac{1}{2}abM \qquad \text{(by § 5.1(7))}. \qquad (1)\end{aligned}$$

On the other hand, the same area may be estimated geometrically as follows.

$$\begin{aligned}(PS\Pi) &= (SQP) + (Q\Pi P) \\ &= \frac{b}{a}(SQP') + \frac{b}{a}(Q\Pi P')\end{aligned}$$

(this last from the affine relation between the ellipse and the circle); therefore,

$$\begin{aligned}(PS\Pi) &= \frac{b}{a}(P'S\Pi) \\ &= \frac{b}{a}(\tfrac{1}{2}a^2E - \tfrac{1}{2}ae\,.\,a\sin E) \\ &= \tfrac{1}{2}ab(E - e\sin E). \qquad (2)\end{aligned}$$

By equating the expressions (1) and (2) we obtain "Kepler's equation"

$$\begin{aligned}E - e\sin E &= M \\ &= n(t - T) \\ &= 2\pi\frac{(t - T)}{P} = 2\pi\frac{t - T}{a^{3/2}}. \qquad (3)\end{aligned}$$

To calculate the position of a planet at the time t, we must first determine the eccentric anomaly E from the known mean anomaly

$$M = \frac{2\pi}{P}(t - T).$$

From E we calculate the true anomaly w, and hence the coordinates ξ and η. We will, therefore, first investigate the dependence of E on M.

One final point, in § 5.1 it was shown that the true anomaly also satisfies an approximate Kepler equation:

$$w - 2e \sin w = M + O(e^2).$$

The coefficient of the eccentricity, however, is doubled.

3. Kepler's equation uniquely determines the eccentric anomaly E for any given eccentricity e and mean anomaly M. Now $E = E(M)$, and it is a function of the following kind. First, we have

$$E = M + \psi(M)$$

where $\psi(M)$, the so-called "equation of the centre" is a one-valued, odd function of M, such that it may be developed in a power series in $(M - M_0)$ providing that $| M - M_0 |$ is sufficiently small. Further, the function $\psi(M)$ is periodic with period 2π, and may be developed in a uniformly and absolutely convergent Fourier series. The quantity $\psi(M)$ is, at most, equal to e.

Some of these assertions follow immediately from differentiation of Kepler's equation $M = E - e \sin E$ with respect to E. This yields

$$\frac{\mathrm{d}M}{\mathrm{d}E} = 1 - e \cos E$$

and so

$$\frac{\mathrm{d}M}{\mathrm{d}E} \geqslant 1 - e > 0.$$

Thus M is a strictly monotonic increasing function of E, and E is thus uniquely determined by M. (This may also be seen from the preceding sub-section.)

Then, further, M is an odd function of E, so, also, E and $\psi(M)$ are odd functions of M.

The one-valued and differentiable function $M = M(E)$ may be developed in a power series in E. The derivative $\mathrm{d}M/\mathrm{d}E$ is non-zero for all real E so, also, $E = E(M)$ may be developed in a power series in $(M - M_0)$ in a sufficiently small region around each M_0.

The periodicity of $\psi = \psi(M)$ may be derived from the equation

$$\begin{aligned}\psi &= e \sin E \\ &= e \sin (M + \psi).\end{aligned}$$

The same value of ψ satisfies this equation if $M + 2\pi$ is written for M. This equation also shows immediately that

$$| \psi(M) | \leqslant e.$$

Further, because the function $\psi(M)$ is periodic and odd, it possesses a Fourier series of the form

$$\sum_{k=1}^{\infty} A_k \sin kM$$

where

$$A_k = \frac{1}{\pi} \int_0^{2\pi} \psi(M) \sin kM \, \mathrm{d}M.$$

Now $\psi(M)$ is analytic, and may, therefore, be differentiated twice. By two successive integrations by parts we obtain

$$\begin{aligned}A_k &= \frac{1}{k} \cdot \frac{1}{\pi} \int_0^{2\pi} \psi'(M) \cos kM \, \mathrm{d}M \\ &= -\frac{1}{k^2} \cdot \frac{1}{\pi} \int_0^{2\pi} \psi''(M) \sin kM \, \mathrm{d}M,\end{aligned}$$

if $\psi'(M)$ and $\psi''(M)$ denote the first and second derivatives of $\psi(M)$ with respect to M. Therefore,

$$A_k = O\left(\frac{1}{k^2}\right),$$

and from this follows the absolute and uniform convergence of the Fourier series towards $\psi(M)$.

Finally, we show how the Fourier coefficients A_k may be evaluated. There is a slight difficulty here because we must integrate the function $\psi(M)$ which is not yet known. We remove this by a transformation of variables, namely

$$\begin{aligned}\psi(M) &= e \sin E, \\ M &= E - e \sin E\end{aligned}$$

and

$$\mathrm{d}M = (1 - e \cos E) \, \mathrm{d}E.$$

So the integral becomes

$$A_k = \frac{1}{\pi}\int_0^{2\pi} \psi(M)\sin kM \,\mathrm{d}M$$

$$= \frac{e}{\pi}\int_0^{2\pi} \sin E \sin[k(E - e\sin E)] \times (1 - e\cos E)\,\mathrm{d}E\,.$$

This may be evaluated, either numerically, or, for example, by developing it in a power series in e. That such a series exists may be seen immediately: the integrand is an analytic function of e; it is even an integral function, that is its series development converges for all values of e. (Usually $e < 1$, or even $e \ll 1$ so that the convergence is excellent.) The integral may therefore be evaluated by integrating term by term. The integration of each single term may be carried out by means of elementary functions—it is only necessary to substitute the sines and cosines of whole number multiples of E for powers of $\sin E$, $\cos E$ and their products, by means of Demoivre's theorem.

For a given M, ψ may also be developed in a power series in e. If $M = 0$, then, $\psi = 0$. If $M \neq 0$ then $e = e(\psi)$ has a non-zero derivative at $\psi = 0$. In the same way as for ψ and M we conclude that $e = e(\psi)$ is an analytic function as well as $\psi = \psi(e)$, and that in a certain neighbourhood around $\psi = 0$ or $e = 0$, respectively, they may be developed in power series. The coefficients of the series for ψ, for instance, are calculated by successive differentiations of the equation

$$\psi = e\sin(M + \psi)$$

at the point $e = 0$. In contrast to the series for the Fourier coefficients, this series does not converge for all e, or even for all $e < 1$. The value of the convergence radius is approximately 2/3. We shall not go into this purely mathematical question, but point out that it was just this question which led Cauchy to undertake his fundamental work on function theory.

4. The actual calculation of the eccentric anomaly $E = E(M)$ is best done by numerically evaluating the quantity

$$\psi = e\sin(M + \psi).$$

We start with the approximate value

$$\psi_0 = 0$$

and then define a series of approximate values $\psi_1, \psi_2, \psi_3 \cdots$ through the recursion formula

$$\psi_{k+1} = e \sin (M + \psi_k).$$

It can easily be shown

(a) that this series converges (and uniformly, for all M),

(b) that it converges to a solution of the above equation for ψ,

(c) that there is only one solution, and

(d) how the error involved in stopping after p iterations may be estimated.

We shall consider this method in some detail because it may be applied to much more general problems and so it is of fundamental importance both theoretically and practically—especially for the differential equations of the third chapter.

We demonstrate (a), the series converges. For,

$$\begin{aligned} | \psi_{k+2} - \psi_{k+1} | &= e \, | \sin (M + \psi_{k+1}) - \sin (M + \psi_k) | \\ &= e \, | \cos (M + \vartheta_k) | \, | \psi_{k+1} - \psi_k | \\ &\leqslant e \, | \psi_{k+1} - \psi_k |. \end{aligned}$$

We have, here, applied the mean value theorem of differential calculus; ϑ_k denotes a number between ψ_k and ψ_{k+1}.

From this estimate it follows that

$$\begin{aligned} | \psi_2 - \psi_1 | &\leqslant e \, | \psi_1 | = e^2 \, | \sin M |, \\ | \psi_3 - \psi_2 | &\leqslant e \, | \psi_2 - \psi_1 | \leqslant e^3 \, | \sin M |, \\ &\cdots\cdots\cdots \\ | \psi_{k+1} - \psi_k | &\leqslant e^{k+1} \, | \sin M |. \end{aligned}$$

Therefore, the series

$$(\psi_1 - \psi_0) + (\psi_2 - \psi_1) + (\psi_3 - \psi_2) + \ldots$$

having the partial sums $\psi_1, \psi_2, \psi_3, \ldots$ converges (uniformly) to a limiting value ψ_*.

We show (b) that ψ_* satisfies the equation

$$\psi = e \sin (M + \psi).$$

For, we have

$$\begin{aligned} | \psi_* - e \sin (M + \psi_*) | &= | \{\psi_* - \psi_{k+1}\} - \\ &\quad - e \{\sin (M + \psi_*) - \sin (M + \psi_k)\} | \\ &\leqslant | \psi_* - \psi_{k+1} | + e \, | \psi_* - \psi_k | \end{aligned}$$

(again, by the mean value theorem), for each $k = 1, 2, \ldots$. Now for $k \to \infty$ the right-hand side tends to zero, and so the left-hand side must vanish also, which proves the theorem.

We show (c) that there is only one solution. Suppose there are two, ψ and $\bar{\psi}$ so that

$$\left.\begin{aligned} \psi &= e \sin (M + \psi), \\ \bar{\psi} &= e \sin (M + \bar{\psi}), \\ \psi &\neq \bar{\psi}. \end{aligned}\right\}$$

Then we would have (again by the mean value theorem)

$$\begin{aligned} | \bar{\psi} - \psi | &= e \, | \sin (M + \bar{\psi}) - \sin (M + \psi) | \\ &\leqslant e \, | \bar{\psi} - \psi |, \end{aligned}$$

For $e < 1$, however, this is only possible if $\bar{\psi} = \psi$, which is contrary to the assumption. There is, therefore, only one solution.

Finally, we estimate (d) the error of the kth approximation ψ_k. We have

$$\begin{aligned} | \psi - \psi_k | &= | (\psi_{k+1} - \psi_k) + (\psi_{k+2} - \psi_{k+1}) + \cdots | \\ &\leqslant | \psi_{k+1} - \psi_k | + | \psi_{k+2} - \psi_{k+1} | + \cdots \\ &\leqslant e^{k+1} | \sin M | + e^{k+2} | \sin M | + \ldots \\ &= \frac{e^{k+1}}{1 - e} | \sin M |. \end{aligned}$$

Thus, for small values of the eccentricity the series converges rapidly.

The convergence can be accelerated by replacing the first approximation $\psi_1 = e \sin M$ by another first approximation closer to the true value ψ (though it might be less convenient for theoretical purposes.) Such an approximation can be found, for instance, in the following way. From

$$\psi = e \sin (M + \psi) \quad \text{and} \quad | \psi | \leqslant 1$$

it follows that

$$\begin{aligned} \psi &= e(\sin M \cos \psi + \cos M \sin \psi) \\ &\approx e(\sin M + \psi \cos M) \end{aligned}$$

or

$$\psi \approx \frac{e \sin M}{1 - e \cos M}.$$

This value is taken as a first approximation from which, by the above iteration, better approximations are readily calculated. (An analogous method can be applied to more general problems; an example will be found in III, § 3.)

5. Now that we know how the eccentric anomaly E can be calculated, we will show how the true anomaly w and the coordinates ξ, η are found. First we consider the determination of w.

We start with the equation

$$\overrightarrow{\mathrm{CS}} + \overrightarrow{\mathrm{SQ}} = \overrightarrow{\mathrm{CQ}} \tag{1}$$

(See Fig. 2.) Clearly

$$\begin{aligned} \overrightarrow{\mathrm{CS}} &= ae, \\ \overrightarrow{\mathrm{SQ}} &= r \cos w \\ &= a \frac{1 - e^2}{1 + e \cos w} \cos w, \\ \overrightarrow{\mathrm{CQ}} &= a \cos E. \end{aligned}$$

If we put these expressions into relation (1) we obtain, by a few intermediate steps

$$\left. \begin{aligned} \cos E &= \frac{e + \cos w}{1 + e \cos w}, \\ \cos w &= \frac{\cos E - e}{1 - e \cos E}. \end{aligned} \right\}$$

Similar sine formulae may be obtained if the ordinate $\overrightarrow{\mathrm{QP}}$ is expressed in terms of E and of w:

$$\begin{aligned} \overrightarrow{\mathrm{QP}} &= \frac{b}{a} \overrightarrow{\mathrm{QP}'} \\ &= \frac{b}{a} a \sin E \\ &= a\sqrt{(1 - e^2)} \sin E, \end{aligned}$$

and

$$\begin{aligned} \overrightarrow{\mathrm{QP}} &= r \sin w \\ &= a \frac{1 - e^2}{1 + e \cos w} \sin w. \end{aligned}$$

Equating these two expressions we obtain

$$\sin E = \sqrt{(1 - e^2)}\,\frac{\sin w}{1 + e \cos w}.$$

To obtain the corresponding formula for $\sin w$, we express $\cos w$ in terms of $\cos E$ and solve for $\sin w$. This gives

$$\sin w = \sqrt{(1 - e^2)}\,\frac{\sin E}{1 - e \cos E}.$$

6. Consider, now, the coordinates

$$\left.\begin{aligned} \xi &= r \cos(\omega + w) = r \cos w \cos \omega - r \sin w \sin \omega, \\ \eta &= r \sin(\omega + w) = r \cos w \sin \omega + r \sin w \cos \omega. \end{aligned}\right\}$$

For convenience in calculation we introduce the coordinates

$$\left.\begin{aligned} \tilde{x} &= r \cos w, \\ \tilde{y} &= r \sin w, \end{aligned}\right\}$$

where the $\tilde{x}$-axis coincides with the major axis of the orbital ellipse. The equations

$$\left.\begin{aligned} \xi &= \tilde{x} \cos \omega - \tilde{y} \sin \omega, \\ \eta &= \tilde{x} \sin \omega + \tilde{y} \sin \omega \end{aligned}\right\}$$

express the fact that the $(\tilde{x}, \tilde{y})$-system has been obtained from the (ξ, η)-system by rotation through an angle ω.

From

$$r = \frac{a(1 - e^2)}{1 + e \cos w} \tag{1}$$

and the formulae of the previous subsection for $\cos w$ and $\sin w$, we obtain, after some intermediate steps

$$r = a(1 - e \cos E), \tag{2}$$

$$\left.\begin{aligned} \tilde{x} &= a(\cos E - e), \\ \tilde{y} &= a\sqrt{(1 - e^2)} \sin E \end{aligned}\right\} \tag{3}$$

$$= b \sin E.$$

Still more simply, these formulae may be obtained directly by a geometrical method. Thus,

$$\left.\begin{aligned} \tilde{x} &= \overrightarrow{SQ} = \overrightarrow{CQ} - \overrightarrow{CS} = a \cos E - ae, \\ \tilde{y} &= \overrightarrow{QP} = \frac{b}{a}\,\overrightarrow{QP'} \qquad = b \sin E, \end{aligned}\right\}$$

and from this we obtain

$$r = \sqrt{(\tilde{x}^2 + \tilde{y}^2)} = a(1 - e \cos E).$$

Further,

$$\left.\begin{aligned} \frac{\xi}{a} &= \cos \omega(\cos E - e) - (1 - e^2)^{1/2} \sin \omega \sin E, \\ \frac{\eta}{a} &= \sin \omega(\cos E - e) + (1 - e^2)^{1/2} \cos \omega \sin E, \end{aligned}\right\}$$

and by § 4.3(2)

$$\left.\begin{aligned} \frac{x}{a} &= (\cos \omega \cos \Omega - \sin \omega \sin \Omega \cos \iota)(\cos E - e) - \\ &\quad -(1 - e^2)^{1/2}(\sin \omega \cos \Omega + \cos \omega \sin \Omega \cos \iota) \sin E, \\ \frac{y}{a} &= (\cos \omega \sin \Omega + \sin \omega \cos \Omega \cos \iota)(\cos E - e) - \\ &\quad -(1 - e^2)^{1/2}(\sin \omega \sin \Omega - \cos \omega \cos \Omega \cos \iota) \sin E, \\ \frac{z}{a} &= \sin \omega \sin \iota(\cos E - e) + (1 - e^2)^{1/2} \cos \omega \sin \iota \sin E. \end{aligned}\right\} \quad (4)$$

In vector form the equations read

$$\boldsymbol{r} = a[\boldsymbol{i}(\cos E - e) + \boldsymbol{j}\sqrt{(1 - e^2)} \sin E] \quad (5)$$

where

$$\boldsymbol{r} = \begin{pmatrix} x \\ y \\ z \end{pmatrix},$$

$$\boldsymbol{i} = \begin{pmatrix} \cos \omega \cos \Omega - \sin \omega \sin \Omega \cos \iota \\ \cos \omega \sin \Omega + \sin \omega \cos \Omega \cos \iota \\ \sin \omega \quad \sin \iota \end{pmatrix} \quad (6)$$

and

$$\boldsymbol{j} = \begin{pmatrix} -\sin \omega \cos \Omega - \cos \omega \sin \Omega \cos \iota \\ -\sin \omega \sin \Omega + \cos \omega \cos \Omega \cos \iota \\ \cos \omega \quad \sin \iota \end{pmatrix}. \quad (7)$$

Thus $\boldsymbol{i}$ is the unit vector in the direction of perihelion, and $\boldsymbol{j}$ is a unit vector in the orbital plane perpendicular to $\boldsymbol{i}$.

7. The problem posed at the beginning of this section was to determine the position of a planet, as a function of time, given its elements. It is solved in the following way: from the mean

anomaly we calculate the eccentric anomaly, and from this we obtain the position vector $\boldsymbol{r}$ by formulae (5)–(7) of the last subsection.

All the coordinates are analytic and periodic functions of the time, and may, therefore, be developed in absolutely and uniformly convergent Fourier series. If it is wished to evaluate these explicitly, it is necessary only to develop the series for $\tilde{x}$ and $\tilde{y}$. Those for any other coordinates may then be obtained from a few of simple algebraic operations. The Fourier coefficients for $\tilde{y}$ are essentially already known; for $\tilde{y}$ is proportional to sin E, and therefore to the function $\psi(M)$, the coefficients of which we have given in § 5.3. Those for $\tilde{x}$ may be obtained in a similar manner.

The eccentric anomaly E is a "uniformizing variable"; that is to say, the coordinates, radius vector and time are one-valued analytic functions of E, whose power series converge for all values of E. Moreover, they may even be represented through elementary closed expressions.

6. Orbital determinations and their improvement

1. We shall now consider the inverse problem, namely the determination of the six elements of a planetary orbit from a number of known positions of the planet. For this we need (at least) six independent equations of condition.

In general, to obtain a unique solution for the problem, it is sufficient to know two positions $\boldsymbol{r}_1$ and $\boldsymbol{r}_2$ at the instants t_1 and t_2 respectively. It is only necessary that these two positions do not lie on the same straight line through the Sun. Together with the Sun they then determine the orbital plane uniquely. Thus, already, we obtain two elements, the inclination ι, and the direction of the nodal line Ω. The calculation is as follows: we define the normal unit vector of the orbital plane $\boldsymbol{k}$ by

$$\boldsymbol{k} = \boldsymbol{i} \times \boldsymbol{j}. \tag{1}$$

From the formulae (6) and (7) of § 5.6, or more simply, from the coordinate transformation § 4.3(2), we obtain, for $\xi = 0$, $\eta = 0$, $\zeta = 1$

$$\boldsymbol{k} = \begin{pmatrix} \sin\Omega\sin\iota \\ -\cos\Omega\sin\iota \\ \cos\iota \end{pmatrix} \tag{2}$$

while, evidently,

$$\boldsymbol{k} = \frac{\boldsymbol{r}_1 \times \boldsymbol{r}_2}{|\boldsymbol{r}_1 \times \boldsymbol{r}_2|}, \tag{3}$$

and hence Ω and ι may be evaluated.

The remaining elements we find in the following way. The equation of the orbital ellipse reads, in polar coordinates,

$$r = a\,\frac{1 - e^2}{1 + e \cos (u - \omega)} \tag{4}$$

where

$$u = \omega + w$$

and is the angle between the position vector of the planet and the nodal direction. In particular, for each of the known positions (r_1, u_1), (r_2, u_2) we have

$$r_\nu = a\,\frac{1 - e^2}{1 + e \cos (u_\nu - \omega)}, \qquad \nu = 1, 2. \tag{5}$$

Thus we have two equations for the three unknowns a, e and ω. A third relation may be obtained from the law of areas, since the area of the sector defined by the two positions must bear the ratio $(t_2 - t_1)/P$ to that of the whole ellipse. The analysis of this condition is difficult and leads to transcendental equations for the unknowns. Therefore, since we make no use of the result, we shall not consider it any further.

If, however, we know a third position of the planet, then the calculation becomes very simple. From the equations (5) for r_1 and r_2, it follows, by eliminating $a(1 - e^2)$, that

$$er_2 \cos (u_2 - \omega) - er_1 \cos (u_1 - \omega) = r_1 - r_2$$

or, applying the addition theorem for cosines and introducing Cartesian coordinates,

$$\xi = r \cos u, \qquad \eta = r \sin u,$$

$$(\xi_2 - \xi_1) \,.\, e \cos \omega + (\eta_2 - \eta_1) \,.\, e \sin \omega = r_1 - r_2 \tag{6}$$

and, similarly,

$$(\xi_3 - \xi_1) \,.\, e \cos \omega + (\eta_3 - \eta_1) \,.\, e \sin \omega = r_1 - r_3.$$

Thus we have two linear equations in the two unknowns $(e \cos \omega)$ and $(e \sin \omega)$. The determinant of the system must be different from zero, otherwise the three points lie on a straight line, not on an ellipse. We can, therefore, solve for $e \cos \omega$ and $e \sin \omega$. From the solution we obtain e and ω and, finally, a is given by one of the equations for r_1, r_2 or r_3. In addition, the law of areas may be used to check the result.

To obtain the time of perihelion the period P is determined from the semi-major axis. Then the eccentric anomaly E is calculated, for any one of the known positions, by means of the formulae

$$\left.\begin{aligned} a(\cos E - e) &= \boldsymbol{r} \cdot \boldsymbol{i} , \\ a(1 - e^2)^{1/2} \sin E &= \boldsymbol{r} \cdot \boldsymbol{j} , \end{aligned}\right\} \tag{7}$$

which follow from § 5.6(5). From this the mean anomaly M is calculated by means of Kepler's equation. The time of perihelion T is then obtained from the relation

$$2\pi \frac{t_i - T}{P} = M_i , \qquad i = 1, 2 \text{ or } 3.$$

In this method, therefore, only one of the three times t_1, t_2, t_3 is used. The other two need not even be known.

2. In the special case when two positions of a planet are "infinitesimally close"—that is when the position vector $\boldsymbol{r}$ and the corresponding velocity vector $\boldsymbol{v}$ are known—the calculation of the elements is simple. We replace $\boldsymbol{r}_1$ by $\boldsymbol{r}$ and $\boldsymbol{r}_2$ by $\boldsymbol{r} + \boldsymbol{v}\, \Delta t + \boldsymbol{O}(\Delta t^2)$, where $\Delta t = t_2 - t_1$. Then equation (3) of § 6.1 gives

$$\boldsymbol{k} = \lim_{\Delta t \to 0} \frac{\boldsymbol{r} \times [\boldsymbol{r} + \boldsymbol{v}\, \Delta t + \boldsymbol{O}(\Delta t^2)]}{|\, \boldsymbol{r} \times [\boldsymbol{r} + \boldsymbol{v}\, \Delta t + \boldsymbol{O}(\Delta t^2)]\,|} = \frac{\boldsymbol{r} \times \boldsymbol{v}}{|\, \boldsymbol{r} \times \boldsymbol{v}\,|}$$

while equation 6.1(6) becomes

$$\dot{\xi} \,.\, e \cos \omega + \dot{\eta} \,.\, e \sin \omega = -\dot{r} \tag{1}$$

where the dot denotes differentiation with respect to the time variable t. The quantities $\dot{\xi}$, $\dot{\eta}$, $\dot{r}$ are, by the assumptions, known. The equation of the ellipse

$$r = \frac{a(1 - e^2)}{1 + e \cos (u - \omega)}$$

we rewrite as

$$r \cos u \,.\, e \cos \omega + r \sin u \,.\, e \sin \omega = a(1 - e^2) - r,$$

that is

$$\xi \,.\, e \cos \omega + \eta \,.\, e \sin \omega = a(1 - e^2) - r. \tag{2}$$

The right-hand side of this equation is known, since, by § 4.4(2) the areal velocity c is given by

$$c = \pi\sqrt{(1 - e^2)} \,.\, \sqrt{a},$$

therefore

$$a\,(1 - e^2) = \left(\frac{c}{\pi}\right)^2$$

and c is also given by

$$c = \tfrac{1}{2}r^2\dot{u} = \tfrac{1}{2}(\xi\dot{\eta} - \eta\dot{\xi}) = \tfrac{1}{2}\,|\,\boldsymbol{r} \times \boldsymbol{v}\,|.$$

The rest of the calculation is as before: the unknowns $e \cos \omega$ and $e \sin \omega$ are evaluated from the linear equations (1) and (2) etc.

3. The problems just discussed have significance in connexion with the practical problem of orbit determination; this is: from a sufficient number of observations of the directions of a planet, as seen from the Earth, and of the corresponding times, to determine the six elements of its orbit.

How many observations of direction are necessary and sufficient for a solution? Each direction is described by two angular coordinates; so, to form the six equations of condition which are needed, three directions at three different instants are necessary. In general, three will also be sufficient.

If, however, the Earth and the planet we are observing move in the same plane, there are then only four elements to determine, namely, the semi-axis, the eccentricity, the direction of perihelion and the time of perihelion. Now, however, each observation yields only one angular coordinate and so, in this case, we need four observations at different times.

In what follows, however, we shall consider briefly only the general case when three observations suffice. We are interested only in the principles, and we sketch a method which, though simple in principle is possibly not very useful in practice.

4. For simplicity we denote the six elements of the orbit of a planet by $a_1, \ldots, a_6$. Further, we assume that approximate values $\mathring{a}_1, \ldots, \mathring{a}_6$ are already known, perhaps from another investigation, or perhaps by plausible inferences. The directions $\boldsymbol{u}_i$ of the planet, as seen from the Earth at the instants t_1, t_2, t_3 are known empirically, and so are the position vectors of the Earth, $\boldsymbol{q}_i$, at the times t_i, $i = 1, 2, 3$. Let $\boldsymbol{r}_i = \boldsymbol{r}(a, t_i)$ be the position vector

of a planet, at time t_i, whose orbital elements are $a_1, a_2, \ldots, a_6$. Similarly, $\mathring{\boldsymbol{r}}_i = \boldsymbol{r}(\mathring{a}, t_i)$. Further, let ϱ_i be the distance of the planet from the Earth at the time t_i, and $\mathring{\varrho}_i$ the approximate distance corresponding to the approximation $\mathring{\boldsymbol{r}}_i$ for the position of the planet.

Then,

$$\boldsymbol{r}_i = \boldsymbol{q}_i + \varrho_i \boldsymbol{u}_i, \tag{1}$$

$$\varrho_i = | \boldsymbol{r}_i - \boldsymbol{q}_i |. \tag{2}$$

Hence, as an approximation of the first order in δa_ν, where

$$\delta a_\nu = a_\nu - \mathring{a}_\nu, \qquad \nu = 1, 2, \ldots, 6,$$

we obtain from (1) that

$$\mathring{\boldsymbol{r}}_i + \sum_{\nu=1}^{6} \frac{\partial \mathring{\boldsymbol{r}}_i}{\partial \mathring{a}_\nu} \delta a_\nu = \boldsymbol{q}_i + (\mathring{\varrho}_i + \delta \varrho_i) \boldsymbol{u}_i, \qquad i = 1, 2, 3, \tag{3}$$

where

$$\delta \varrho_i = \varrho_i - \mathring{\varrho}_i, \qquad i = 1, 2, 3.$$

Again, to the first order in δa_ν,

$$\begin{aligned} \delta \varrho_i &= \sum_{\nu=1}^{6} \frac{\partial \mathring{\varrho}_i}{\partial \mathring{a}_\nu} \delta a_\nu = \sum_{\nu=1}^{6} \left(\frac{\partial \mathring{\varrho}_i}{\partial \mathring{x}_i} \frac{\partial \mathring{x}_i}{\partial \mathring{a}_\nu} + \frac{\partial \mathring{\varrho}_i}{\partial \mathring{y}_i} \frac{\partial \mathring{y}_i}{\partial \mathring{a}_i} + \frac{\partial \mathring{\varrho}_i}{\partial \mathring{z}_i} \frac{\partial \mathring{z}_i}{\partial \mathring{a}_\nu} \right) \delta a_\nu \\ &= \sum_{\nu=1}^{6} \frac{\mathring{\boldsymbol{r}}_i - \boldsymbol{q}_i}{| \mathring{\boldsymbol{r}}_i - \boldsymbol{q}_i |} \frac{\partial \mathring{\boldsymbol{r}}_i}{\partial \mathring{a}_\nu} \delta a_\nu \\ &= \sum_{\nu=1}^{6} \frac{\boldsymbol{r}_i - \boldsymbol{q}_i}{| \boldsymbol{r}_i - \boldsymbol{q}_i |} \frac{\partial \mathring{\boldsymbol{r}}_i}{\partial a_\nu} \delta a_\nu, \end{aligned}$$

second order terms once more being neglected. Hence by (1),

$$\delta \varrho_i = \boldsymbol{u}_i \sum_{\nu=1}^{6} \frac{\partial \mathring{\boldsymbol{r}}_i}{\partial a_\nu} \delta a_\nu. \tag{4}$$

Now, for each value of i, equations (3) contain three scalar equations. These equations are not independent, however, for

$$\boldsymbol{u}_i^2 = \left(\frac{\boldsymbol{r}_i - \boldsymbol{q}_i}{\varrho_i} \right)^2 = 1, \qquad i = 1, 2, 3.$$

Therefore, one equation of each triplet corresponding to each value of i may be omitted. (It is best to omit the equations relating to the z-components, since all these components take only small values and are, therefore, the least sensitive to variations in the elements. Conversely, the determination of the elements from the z-components is the least accurate.)

By substituting the expressions (4) into equations (3) (omitting those in z) we obtain six equations

$$\sum_{\nu=1}^{6}\left[\frac{\partial \mathring{\boldsymbol{r}}_i}{\partial \mathring{a}_\nu} - \left(\frac{\partial \mathring{\boldsymbol{r}}_i}{\partial \mathring{a}_\nu}\,\boldsymbol{u}_i\right)\boldsymbol{u}_i\right]\delta a_\nu = \boldsymbol{q}_i + \mathring{\varrho}_i \boldsymbol{u}_i - \mathring{\boldsymbol{r}}_i, \quad i = 1, 2, 3, \quad (5)$$

for the six unknowns δa_ν. Here $\mathring{\boldsymbol{r}}_i$ is given by equation § 5.6(5). In differentiating, the eccentric anomaly E must also be regarded as a function of a, E and T. Thus, from Kepler's equation

$$E - e \sin E = 2\pi \frac{t - T}{a^{3/2}},$$

it follows that

$$\begin{cases} (1 - e \cos E)\dfrac{\partial E}{\partial a} = -3\pi \dfrac{t - T}{a^{5/2}}, \\[2ex] (1 - e \cos E)\dfrac{\partial E}{\partial e} = \sin E, \\[2ex] (1 - e \cos E)\dfrac{\partial E}{\partial T} = -\dfrac{2\pi}{a^{3/2}}. \end{cases}$$

From the equation (5) we can calculate the improvements δa_ν. From these we obtain the new approximate solution $\mathring{a}_\nu + \delta a_\nu$ and we repeat the calculation with these, and continue until two successive approximations differ by a sufficiently small amount. Obviously this method is a generalization of Newton's method of solving an equation with only one unknown.

So far, we have not developed a method for obtaining the first approximation. This problem is purely kinematical, so that it is possible in principle and, indeed, from the point of view of methodology, desirable to develop it now. In our presentation, however, we prefer to postpone this development to II § 4. The work needed for this method will be found in II § 2.1, § 2.4 and § 3.1.

5. The method expounded in § 4 also gives the basic relations for the correction of elements which are approximately known. Let there be given a set of approximations $\mathring{a}_\nu$ to the elements a_ν of a planet, and let the approximation be of such a quality that

the position vector $\boldsymbol{r}(a, t)$ can be represented, with sufficient accuracy by $\boldsymbol{r}(\mathring{a}, t) + \sum_{\nu=1}^{6} \frac{\partial \mathring{\boldsymbol{r}}}{\partial \mathring{a}_\nu} \delta a_\nu$. Further, let us assume that the directions $\boldsymbol{u}_i$ of the planet, as seen from the Earth, are known at the times $t_1, t_2, \ldots, t_n$, where $n > 3$. The problem is to determine corrections δa_ν of the elements $\mathring{a}_\nu$, which take account of the observations $\boldsymbol{u}_i$.

If the measurements made at the times $t_1, \ldots, t_n$ were absolutely precise, the corrections δa_ν could be determined from any three of the vectors $\boldsymbol{u}_i$. In practice, however, there are errors in the measurements, so that the values of δa_ν derived from different sets of vectors are, in general, different from each other. In other words, the equations 4(5), which we abbreviate to the form

$$\sum_{\nu=1}^{6} \boldsymbol{f}_\nu(\mathring{a}, \boldsymbol{u}_i, t_i)\, \delta a_\nu - \boldsymbol{g}(\mathring{a}, \boldsymbol{u}_i, t_i) = 0, \qquad i = 1, 2, \ldots, n, \tag{1}$$

are, in general, incompatible. (It should be remembered that we use only the x- and y-components of these equations. In what follows, it will be convenient to replace the third components by zero.)

We now take the minimum of the quadratic form

$$Q = \sum_{i=1}^{n} \left[\sum_{\nu=1}^{6} \boldsymbol{f}_\nu(\mathring{a}, \boldsymbol{u}_i, t_i)\lambda_\nu - \boldsymbol{g}(\mathring{a}, \boldsymbol{u}_i, t_i) \right]^2$$

as a measure of the incompatibility of equations (1). The λ_ν's denote real variables in the range $-\infty \ldots +\infty$. Evidently, this measure of incompatibility vanishes if and only if the equations are compatible. In all other cases it is positive. The values of λ_ν for which Q is smallest are regarded as the most plausible values of δa_ν. Thus, by differentiating Q with respect to λ_ν, the following system of linear equations for δa_ν is obtained:

$$\sum_{i=1}^{n} \left[\sum_{\nu=1}^{6} \boldsymbol{f}_\nu(\mathring{a}, \boldsymbol{u}_i, t_i)\, \delta a_\nu - \boldsymbol{g}(\mathring{a}, \boldsymbol{u}_i, t_i) \right] \boldsymbol{f}_\nu(\mathring{a}, \boldsymbol{u}_i, t_i) = 0,$$

or

$$\sum_{\nu=1}^{6} \left[\sum_{i=1}^{n} \boldsymbol{f}_\nu^2(\mathring{a}, \boldsymbol{u}_i, t_i) \right] \delta a_\nu = \sum_{\nu=1}^{6} \boldsymbol{f}_\nu(\mathring{a}, \boldsymbol{u}_i, t_i) \boldsymbol{g}_\nu(\mathring{a}, \boldsymbol{u}_i, t_i).$$

In practice this system always has a uniquely determined solution. The corrected elements are then $\mathring{a}_\nu + \delta a_\nu$.

If the assumed linear approximation is not sufficiently good, either we can apply an iterative method, or else we can consider higher order terms. We shall not, however, consider this any further.

6. It must be pointed out that some of the elements depend very sensitively on the observational data, so that quite small observational errors can produce much larger errors in these particular elements.

First, let us consider the orbital plane of the planet. In general it intersects that of the Earth at a very small angle. Small uncertainties in the observations give small errors in the orientation of the orbital plane, if we describe it by its normal vector. Small variations in the direction of the normal vector, however, can produce large variations in the direction of the nodal line. The value of Ω extracted from the observations is, therefore, relatively uncertain.

Similar considerations hold for the direction of perihelion. Planetary orbits are nearly circular, so that observational errors can lead to a completely false direction of perihelion. In the same way the eccentricity is relatively uncertain. Suppose, for instance, that the eccentricity and direction of perihelion are determined from $e \cos \omega$ and $e \sin \omega$, (as in § 6.1) these latter quantities having been determined empirically. Let Δc and Δs be their observational errors, and Δe and $\Delta\omega$ be the corresponding errors of e and ω. Then,

$$\left.\begin{aligned} \Delta(e \cos \omega) &= \Delta c, \\ \Delta(e \sin \omega) &= \Delta s, \end{aligned}\right\}$$

in the first approximation,

$$\left.\begin{aligned} \Delta e \cos \omega - e \sin \omega \, \Delta\omega &= \Delta c, \\ \Delta e \sin \omega + e \cos \omega \, \Delta\omega &= \Delta s; \end{aligned}\right\}$$

therefore

$$\left.\begin{aligned} \frac{\Delta e}{e} &= \frac{\Delta c}{e} \cos \omega + \frac{\Delta s}{e} \sin \omega, \\ \Delta\omega &= -\frac{\Delta c}{e} \sin \omega + \frac{\Delta s}{e} \cos \omega. \end{aligned}\right\}$$

Since e is small, the right-hand sides of these equations may become large, because the errors Δc and Δs are given by the observations and cannot be made arbitrarily small.

The large relative error in the eccentricity is, of course, without significance in practice. The absolute error is very small, since the eccentricity itself is very small. The directions of the node and of perihelion, on the other hand, react very sensitively under certain circumstances to small observational uncertainties. Conversely, the coordinates are insensitive to uncertainties in these elements so that the uncertainty of a predicted position is not appreciably greater than that of an observation.

7. Hitherto we have assumed the orbit of the Earth to be known. It can, of course, be determined empirically by observing the apparent annual motion of the Sun. Suppose, however, we wish to determine the elements of the Earth's orbit from only a few observations, how would we proceed?

First, we have at once, by definition, the semi-major axis $a = 1$. The orientation of the orbital plane with respect to the fixed stars may be determined from two observations of the Sun. The direction and time of perihelion and the eccentricity can be determined from the dependence between the true and mean anomalies. These read, in the first approximation:

$$w - 2e \,.\, \sin w = M + O(e^2).$$

For w we substitute, as in § 6.1, $u - \omega$ where u denotes the observed direction of any fixed zero direction (e.g. the "vernal equinox"). The unknowns are e, ω and M, the last on account of the unknown time of perihelion T. We have, therefore, three unknowns, e, ω, and T. Correspondingly we set up three equations

$$u_i - \omega - 2e \sin (u_i - \omega) = 2\pi(t_i - T) + O(e^2), \qquad i = 1, 2, 3.$$

We subtract the first from the second and third, and then use the addition theorem for sines. This gives two linear equations for the unknowns $e \sin \omega$ and $e \cos \omega$, the terms in ω and T being eliminated. From these, in turn, $e \cos \omega$, $e \sin \omega$, e, ω and T may be calculated, if observations at three instants, t_i, are available.

The method is similar to that for determining the orbital elements from three observations of a planet (§ 6.1). It yields only an approximation; but, obviously, the approximation may include as many higher terms as desired, although the computation would then become rather complicated.

8. In comparing the computed and observed positions of a planet, we must also consider the finite velocity of light (300,000 km sec^{-1}). The observed position of a planet is not its position at the instant of observation, but the position it had at some previous instant. The difference in the two times is the distance of the planet divided by the velocity of light. The time data in the various formulae must, therefore, be corrected for this so-called aberration time, if they are to be applied to the observations (Cf. Exercises 23–25).

7. Summary and discussion

In this first chapter we have discussed the kinematics of a single planet. We have made clear how Kepler's laws may be found empirically, using his own method. Then we applied these to the calculation of planetary orbits as well as to the determination of orbital elements from only a few observations.

Such a purely descriptive theory may be developed in either of two directions. First, the motion of the moons around their planets, and of the double stars, can be investigated in a similar way, and the results compared with those of this chapter. Secondly, recalling the remarks of § 4.1, we remember that Kepler's laws hold only approximately. We could, therefore, try to refine them by purely empirical means, for example, by the following method: we determine the Kepler ellipse by the method of determining orbits from three observations, observe the variations of the orbital elements over long periods of time, and compare these variations with the neighbouring planets. In this way we might arrive at some quantitative law. The elements could be introduced into the formulae as (curtailed) Fourier or power series of the time variable.

Kepler saw another way without, however, being able to follow it himself. The fact that the Sun occupies a privileged place in each planetary orbit, namely one of the foci, led him to postulate in it the origin of a dynamical "force". He came near to the truth since, at times, he thought that this force was propagated according to the same law as for light. He was not able to clarify his concepts, however. This was done, at least partially, by Newton. (The principles and the method were clear to him, but not all the fundamental concepts.)

Although all the possibilities of a purely phenomenological

approach are not completely exhausted, we shall also abandon the kinematical approach and turn to dynamics. Our aim is to gain a deeper insight. This can only be done at the cost of introducing hypotheses. The determination and calculation of an orbit is essentially a kinematical problem, and, therefore, its validity does not depend on that of the basic dynamical hypotheses.

Exercises

1. Assuming that the Earth and a planet move around the Sun in the same plane, and with uniform circular motion, construct diagrams showing the direction of the planet, as seen from the Earth, as a function of the time,
   (*a*) for an inner planet ($a < 1$),
   (*b*) for an outer planet ($a > 1$).
2. Derive the formula for the sidereal period assuming that the Earth, and the projection of Mars on the Earth's orbital plane move around the Sun with constant angular velocities (§ 1.4).
3. Does the derivation of the relation between the sidereal and synodic periods actually use all the assumptions (i)–(iv)?
4. Give a formula for the relative error in the semi-major axis caused by the observational error of the synodic period (§§ 1.4 and 3.2).
5. Give a rigorous proof of the formula (1) of § 1.5.
6. Describe, in full detail, how the direction Sun–Earth may be empirically determined as a function of time (§ 2.1).
7. State the formulae for Kepler's construction in a plane (§ 2).
8. Describe, in detail, Kepler's construction in three dimensions (§ 2).
9. Can the construction of the last question also be applied to the inner planets (§ 2.3)?
10. Derive the formula for the areal velocity used in § 5.1.
11. Prove that the different formulations of Kepler's second law are equivalent (§ 3.1).
12. Prove the transformation formulae given in § 4.3, following the instructions given there.
13. Derive the equation for an ellipse in polar coordinates as given in § 5.1.
14. Continue the development of w in terms of e, given in § 5.1, one term further.
15. Evaluate the integral (11), of § 5.1, by substituting $s = \tan \frac{1}{2} w$.
16. Develop the Fourier coefficients, A_k, in powers of e, up to and including terms of the second order (§ 5.3).
17. Give an exact estimate for the error of the linear approximation
$$A_k \approx A_k^{(0)} + eA_k^{(1)}.$$
18. Show, by successive integrations by parts, (as in the example given in § 5.3) that for each natural number m
$$| A_k | \leqslant \frac{C(m)}{k^m}, \qquad k = 1, 2, 3, \ldots$$

where $C(m)$ is a positive number depending only on m. (Hence, you may conclude that the Fourier series for $\psi(M)$ converges quickly. By means of the theory of complex functions it may easily be shown to converge asymptotically like a geometric series.)

19. The square of the mean error of $\sum_{k=1}^{n} A_k \sin kM$ as an approximation to $\psi(M)$ is given by

$$\frac{1}{2\pi}\int_0^{2\pi}\left\{\psi(M) - \sum_1^n A_k \sin kM\right\}^2 \mathrm{d}M$$
$$= \frac{1}{2\pi}\int_0^{2\pi}\{\psi(M)\}^2 \,\mathrm{d}M - \tfrac{1}{2}\sum_1^n A_k^2.$$

Calculate $\int_0^{2\pi}\{\psi(M)\}^2 \,\mathrm{d}M$ by the method of § 5.3, remembering that $\psi = e \sin E$.

20. For readers with some knowledge of function theory. The eccentric anomaly E can be developed in a power series in e. (§ 5.3). Determine its convergence radius as the distance from the point $e = 0$ to the next singularity. (Consider, here, only that branch of the function $E \equiv E(e)$ which assumes real values on the real axis in the vicinity of the null point.)

21. After how many iterations is

$$|\psi - \psi_k| < 1''$$

if e is assumed to be the eccentricity of the orbit of the Earth (§ 5.4)?

22. Give a geometrical interpretation of equations (5) of § 6.4.

23. How must the aberration time be taken into account when determining orbits (§ 6.8)?

24. In what way and what amount does the aberration time change in the course of a year? In 1675 Olaf Römer measured this variation, by noting the times of the eclipses of the moons of Jupiter by the shadow of the planet. From this he determined the velocity of light. Describe his method in full detail (§ 6.8).

25. Conversely, how can the absolute linear dimensions of the planetary system be determined from the aberration time of Jupiter and laboratory measurements of the velocity of light (§ 6.8)?

26. Give a concise outline of all that must be done to determine the absolute linear dimensions of the planetary system (§ 3.3).

NOTES

§ 1.2. Cf.

[1] Pilowski, K., Bemerkungen zu den Grundlagen der sphärischen Astronomie, *Astr. Nachr.* **281**, 53, 1953.

[2] Kurth, R., On the Foundations of Celestial Mechanics. (Will probably be published in *Astr. Nachr.*)

It is not necessary to assume the directions of the "fundamental stars" to be fixed. There is no objection in principle to attributing an arbitrary rotation to the coordinate system, with respect to these stars but, of course, such a system is of no practical value.

§ **1.3.** For descriptions of the motion of the planets as seen from the Earth, cf., for example,

[3] BECKER, F., *Grundriss der sphärischen und der praktischen Astronomie.*

[4] SMART, W. M., *Spherical Astronomy*, Cambridge, 1949.

§ **1.5.** Observational data quoted in the text are mostly taken from

[5] ALLEN, C. W., *Astrophysical Quantities*, London, 1955.

§ **2.2–3** and § **3.2.** Kepler published his first and second laws in his *Astronomia Nova* (1603). He had already considered the problem of the connection between periods and semi-major axes in his first work *Mysterium Cosmographicum* (1595), but he did not solve it until 1618 and he published his third law in his work *De Harmonice Mundi* (1619).

§ **3.3.** The orbital velocity of the Earth can also be determined, in a similar way, from the aberration of the fixed stars. Cf. [3] or

[6] PLUMMER, H. C., *An Introductory Treatise on Dynamical Astronomy*, Cambridge, 1918.

§ **5.3.** The coefficients A_k are closely related to Bessel's functions. Cf. [6] or

[7] WINTNER, A., *The Analytical Foundations of Celestial Mechanics*, Princeton, 1947.

§ **6.3.** For information on the determination of orbits, cf., for example,

[8] BAUSCHINGER, J., *Die Bahnbestimmung der Himmelskörper*, Leipzig, 1928.

[9] STRACKE, G., *Bahnbestimmung der Planeten und Kometen*, Berlin, 1929.

The last book deals with the problem solely from the viewpoint of practical computation.

§ **6.5.** In the literature the statement is sometimes made that the application of the method of least squares supposes the validity of Gauss' law of errors. This, however, is not correct. The text follows an idea of Legendre. The foundation of the method on the theory of probability is, in the author's opinion, not more profound, but only more complicated. Cf.

[10] WHITTAKER, E. and ROBINSON, G., *The Calculus of Observations*, London, 1952.

§ **7.** Suppose that much more accurate observations had been available to Kepler. The development of the coordinates, or the elements, as functions of the time, in Fourier series would have appeared to him as follows. The general cosine term (say) has the form

$$A_{\nu\nu_1\nu_2\ldots} \cos \{(\nu n + \nu_1 n_1 + \nu_2 n_2 + \ldots)t\},$$

where n is the mean motion of the planet being considered, and $n_1, n_2 \ldots$ those of neighbouring planets; $\nu, \nu_1, \nu_2 \ldots$ are any integers and the $A_{\nu\nu_1\nu_2\ldots}$ are constants (see III § 2.2). The mean motions are, therefore, related by a sort of linear combination principle with whole number coefficients. Kepler, highly pleased with the whole numbers, would presumably have concluded from this that the mean motions are the fundamental quantities, and would have built his theory on them. Observations of, for example, the satellites of Jupiter, in which a linear relation with (absolutely small) whole number coefficients holds, or of the commensurabilities amongst the Minor Planets, or in the rings of Saturn, would have confirmed the theory. Moreover, the mean motion n is related to the semi-major axis a by $n = 2\pi a^{-3/2}$. Since, for the major planets, a can only take discrete values determined by some law, n also possesses discrete "levels." Similar considerations hold in the satellite systems. (As for the exceptions, suitable rules would have been found for them.) For certain coefficients $A_{\nu\nu_1\nu_2\ldots}$ (i.e. for certain values of ν, ν_1, ν_2 ...) there are "selection rules", for others "resonance rules" (see III § 2.2) etc. Doubtless a theory of the motions of the heavenly bodies may be constructed in this way, that differs radically from the dynamical celestial mechanics, yet is simple, empirically correct and fruitful. Kepler could have developed this, had he had better observational techniques at his disposal. Had he done so, physical science and philosophy would have developed in a completely different way.

CHAPTER II

THE DYNAMICS OF A SINGLE PLANET

1. The law of gravity and the law of motion

1. The everyday meaning of the word "force" is so vague, that it can scarcely be made to yield a satisfactory general intuition. Doubtless both Newton and Kepler remembered this everyday concept when they spoke of physical force. Indeed, this analogy is still plausible, even for us. In order to make it fruitful in celestial mechanics, we shall abstract from this everyday concept a series of postulates which must hold if any idea of force is to be scientifically useful. In what follows, we shall speak of force only in terms of these postulates. Thus we hope to obtain some formal clarity at least.

To this end we must first introduce the idea of a mass point. We use this term to mean a massive body whose diameter is negligibly small compared with its distances from all other bodies, and whose translatory motions are unaffected by any possible rotation. In general, we shall consider the celestial bodies as mass points.

Our first postulates for the concept of force are, then:

First: If one mass point exerts a force (either attractive or repulsive) on another, this force acts along the line joining the two points.

Second: For given mass points the magnitude of the force depends only on their distance apart.

Third: All mass points (all celestial bodies) exert an attractive force on each other (the force of gravitation).

By the third postulate, not only does the Sun exercise an attractive force, for example, on Mars, but so also do all the other planets. The position of the Sun is exceptional, however, and we conclude that its force so outweighs those of all the other planets, that (to the standard of accuracy assumed in the first chapter) the total force is equal to that of the Sun. To this accuracy, the interactions between the planets are negligibly small.

2. We now consider how to define the motive force more

closely. The method by which we search for an answer is clear. We have to find a kinematically defined vector which satisfies the three force postulates. The variety of such vectors may be reduced by adding further plausible postulates, until only one vector satisfies all the postulates. This one will be defined as the force vector.

For Kepler motion the velocity vector does not satisfy the second postulate and it may, therefore, be eliminated immediately. The vector of relative acceleration, however, fulfils all three postulates, as we will now show.

From

$$\boldsymbol{r} = a[\boldsymbol{i}(\cos E - e) + \boldsymbol{j}\sqrt{(1 - e^2)} \sin E] \tag{1}$$

(I, § 5.6(5)) it follows, by differentiating with respect to the time variable t, that

$$\dot{\boldsymbol{r}} = a[-\boldsymbol{i} \sin E + \boldsymbol{j}\sqrt{(1 - e^2)} \cos E)\dot{E},$$

where dots denote differentiation with respect to the time variable. We can obtain an expression for $\dot{E}$ by differentiating Kepler's equation (I, § 5.2(3)) with respect to time. Thus, from

$$E - e \sin E = 2\pi \frac{(t - T)}{a^{3/2}},$$

we obtain

$$(1 - e \cos E)\dot{E} = \frac{2\pi}{a^{3/2}},$$

so that

$$\dot{\boldsymbol{r}} = \frac{2\pi}{\sqrt{a}} (1 - e \cos E)^{-1}[-\boldsymbol{i} \sin E + \boldsymbol{j}\sqrt{(1 - e^2)} \cos E]. \tag{2}$$

From this, differentiating again with respect to t and substituting for $\dot{E}$, we obtain

$$\ddot{\boldsymbol{r}} = -\frac{4\pi^2}{a^2} [\boldsymbol{i}(\cos E - e) + \boldsymbol{j}\sqrt{(1 - e^2)} \sin E](1 - e \cos E)^{-3}.$$

If we substitute, on the right-hand side of this equation, the formula (I § 5.6(2)) for the radius vector r

$$r = a(1 - e \cos E),$$

so we obtain, finally,

$$\ddot{\boldsymbol{r}} = -\frac{4\pi^2}{r^2} \cdot \frac{\boldsymbol{r}}{r}.$$

The vector of acceleration, therefore, is always directed toward the Sun and its magnitude

$$| \ddot{\boldsymbol{r}}(t) | = \frac{4\pi^2}{[r(t)]^2}$$

depends only on the instantaneous radius vector.

The same holds for any vector $\alpha\ddot{\boldsymbol{r}}(t)$ (where α is a positive constant) and for all the known planets. Therefore $\alpha\ddot{\boldsymbol{r}}(t)$ also satisfies the force postulates of subsection 1. Now, by induction, we assume that the force exerted by *any* one mass point upon any other is directed towards it and has a magnitude proportional to $4\pi^2/r^2$, where r is the distance between the two mass points. Thus we still leave open the possibility that the factor of proportionality is different for different mass points. For the planets, however, the law of motion reads: the position vector of any planet is a solution of the differential equation

$$\ddot{\boldsymbol{r}} = -\frac{4\pi^2}{r^2}\frac{\boldsymbol{r}}{r}.$$

That is our preliminary formulation of the Newtonian laws of motion and of gravity. It still requires an essential correction which is, however, below the limits of accuracy hitherto adopted. We shall give this correction in 4. First, however, two critical remarks must be made about the foregoing reasoning.

3. The first remark concerns the way in which we arrived at the law of gravity. We differentiated the coordinates of the planet twice with respect to time. To express these coordinates as functions of time, we used fairly simple functions but, after all, they are empirically determined functions. These are, in principle, known only with a limited accuracy. In fact, we have, in the words of Felix Klein, not functions but "function strips." These "function strips", however, will not give reliable results when differentiated—and certainly not when differentiated twice.

The apparent guarantee of the validity of the law of gravity provided by the formal rigour of the applied mathematics is deceptive. Because of the differentiation the uncertainty of the law may be considerably greater than that of the observations. (The inductive generalization introduces another source of uncertainty.) We shall, therefore, consider Newton's law as an hypothesis and we shall not be unduly surprised if it fails to agree with particular observations. A very small disagreement with observation is known to exist in the motion of Mercury (see III § 7.) Up till now, however, the hypothesis (in the improved formulation, still to be given) has, in general, stood the test very well.

It is nonetheless noteworthy that the dependence on distance is exactly of the form $1/r^2$. It is as if force were a substance spreading uniformly, from the mass point, over concentric spherical surfaces. In earlier times it was thought to be possible to obtain Newton's law of gravity deductively from this idea. Such attempts, however, are open to serious criticism.

Our second note concerns the result of our derivation. In 2 we said we would consider all the kinematic vectors which satisfy the three force postulates of 1. Actually we have only considered vectors proportional to the acceleration. There still remains the question whether there are, perhaps, other vectors to be considered. We do not know the answer, for the question seems never to have been investigated. It would seem improbable that there are any such vectors. If there were, then the law of motion would be correspondingly altered, and the old and questionable new laws of motion would yield the same results for the planetary motions. However, perhaps other forms of motion exist which could, and must, show up the differences between the various laws of motion. We shall return to this point once more, in the following subsection.

4. The law of motion of § 1.2,

$$\ddot{\boldsymbol{r}} = -\frac{4\pi^2}{r^2}\frac{\boldsymbol{r}}{r}$$

gives the acceleration of a planet relative to the Sun. This law must be altered in some way if we wish to think in terms of a general gravitation. For, at the beginning of our investigation, we assumed that the interactions of the planets were much smaller than the action of the Sun on the planets. The empirical law of gravity must be modified to represent these interactions. Now, in § 1.2 we said that the accelerating force is proportional to the vector $\boldsymbol{r}/r^3$, and that the constant of proportionality depends on the pair of mass points considered. We suppose, therefore, that for any two gravitating mass points with position vectors $\boldsymbol{r}_1$ and $\boldsymbol{r}_2$, the acceleration of point 2, relative to point 1, is given by

$$(\boldsymbol{r}_2 - \boldsymbol{r}_1)^{\cdot\cdot} = -\frac{\alpha_{12}}{|\boldsymbol{r}_2 - \boldsymbol{r}_1|} \cdot \frac{\boldsymbol{r}_2 - \boldsymbol{r}_1}{|\boldsymbol{r}_2 - \boldsymbol{r}_1|}, \quad \alpha_{12} = \text{const} > 0. \quad (1)$$

The relative acceleration of point 1 with respect to point 2 is equal and opposite to this. Clearly, $\alpha_{12} = \alpha_{21}$—the constant of proportionality is symmetrical with respect to the two mass points.

We suppose, as a hypothesis, that each mass point has a positive constant, m_1 associated with it in such a way that for any two gravitating points 1 and 2

$$\alpha_{12} = \alpha_{21} = G\,.\,(m_1 + m_2),$$

where G denotes a universal constant whose value depends on the system of units used. This starting point recommends itself by its particular simplicity. The quantities m_1, m_2 are called the masses of the respective particles. The universal constant G is called the constant of gravitation. Equation (1), therefore, takes the, still hypothetical, form:

$$(\boldsymbol{r}_2 - \boldsymbol{r}_1)^{\cdot\cdot} = -G\frac{m_1 + m_2}{|\boldsymbol{r}_2 - \boldsymbol{r}_1|^2} \cdot \frac{\boldsymbol{r}_2 - \boldsymbol{r}_1}{|\boldsymbol{r}_2 - \boldsymbol{r}_1|}. \quad (2)$$

This equation can now be interpreted in such a way as to assign a definite share of the relative acceleration to each particle. Such a representation is based on the idea of "absolute space". Thus it could perhaps be argued that if two mass points attract each other reciprocally, not only one, or the other of them is accelerated, but they are both accelerated, in the inverse ratio to the quantity of matter each contains which is measured by the masses. If, therefore, the coordinate system is at rest in the absolute space, we may write

$$\left.\begin{aligned} \ddot{\boldsymbol{r}}_1 &= Gm_2\frac{\boldsymbol{r}_2 - \boldsymbol{r}_1}{|\boldsymbol{r}_2 - \boldsymbol{r}_1|^3}, \\ \ddot{\boldsymbol{r}}_2 &= Gm_1\frac{\boldsymbol{r}_1 - \boldsymbol{r}_2}{|\boldsymbol{r}_1 - \boldsymbol{r}_2|^3}. \end{aligned}\right\}$$

A coordinate system absolutely at rest cannot, of course, be realized. It is used here only as an auxiliary hypothesis which can, and must, be justified only by its fruitfulness.

Our first hypothesis for the α_{ij}'s can, in principle, be justified by investigating a system of n gravitating mass points, where $n \geqslant 4$. For from n points, $\binom{n}{2} = \frac{1}{2}n(n-1)$ different pairs of points may be selected, so that we shall have $\binom{n}{2}$ distinct α_{ij}'s. Now $\binom{n}{2} > n$ if $n > 3$. If the hypothesis is correct, then any n of the α_{ij}'s determine the other $\binom{n}{2} - n$ uniquely, and the theoretical and empirical values must agree.

The magnitude of the gravitational force we have not yet given. To do so, we must state a further general force postulate which, like the earlier ones, we abstract from everyday experience.

Fourth Force Postulate: The force $\boldsymbol{f}_{12}$ that a mass point 1 exerts on a mass point 2 is equal and opposite to the force $\boldsymbol{f}_{21}$ exerted by the point 2 on the point 1. That is,

$$\boldsymbol{f}_{12} = -\boldsymbol{f}_{21}.$$

In the case of gravitating mass points, we know from the equations for the vectors $\ddot{\boldsymbol{r}}_1$ and $\ddot{\boldsymbol{r}}_2$ (which are proportional to the vectors $\boldsymbol{f}_1$ and $\boldsymbol{f}_2$) that this postulate is fulfilled if

$$\boldsymbol{f}_{21} = CGm_1m_2 \frac{\boldsymbol{r}_2 - \boldsymbol{r}_1}{|\boldsymbol{r}_2 - \boldsymbol{r}_1|^3},$$

where C is a positive, but otherwise arbitrary, universal constant. We simply put it equal to one and so we specify the unit of force and give the equations of motion the form

$$\left.\begin{aligned} m_1\ddot{\boldsymbol{r}}_1 &= \boldsymbol{f}_{21}, \\ m_2\ddot{\boldsymbol{r}}_2 &= \boldsymbol{f}_{12}, \end{aligned}\right\}$$

where

$$\boldsymbol{f}_{21} = Gm_1m_2 \frac{\boldsymbol{r}_2 - \boldsymbol{r}_1}{|\boldsymbol{r}_2 - \boldsymbol{r}_1|^3}.$$

That is Newton's law of gravitation. The leading thought in its original statement was obviously the proportionality of the force to the mass of the particle, where the mass was defined by Newton as the "quantity of matter". The two equations for the accelerations express Newton's law of motion:

$$\text{mass} \times \text{absolute acceleration} = \text{force}.$$

It has been inductively generalized into a basic law of the whole of dynamics. Its many tests, in a variety of applications, show the law to be an excellent approximation to the true state of affairs. The hypothetical objection at the end of § 1.3 is, therefore, of no relevance to reality.

5. We now summarize the fundamental hypotheses.

I. If a freely movable mass m is acted on by a force $\boldsymbol{f}$, then it receives an "absolute acceleration" given by

$$m\ddot{\boldsymbol{r}} = \boldsymbol{f}.$$

II. If there are two gravitating particles, then the point 2 acts on the point 1 with a force $\boldsymbol{f}_{21}$ given by

$$\boldsymbol{f}_{21} = Gm_1m_2 \frac{\boldsymbol{r}_2 - \boldsymbol{r}_1}{|\boldsymbol{r}_2 - \boldsymbol{r}_1|^3}.$$

6. In conclusion we may say a few words about the constant of gravitation, and also about the masses of the Sun and planets. By the general law of gravitation the planets must attract each other. Obviously, however, the gravitation of the Sun so strongly outweighs that of the planets that, to the degree of observational accuracy which we are at present accepting, the interactions of the planets are still negligibly small. Because their distances from each other, at least at times, are not greater than their distances from the Sun, the smallness of the interplanetary forces must arise from the small masses of the planets (relative to the Sun) rather than from their great distances.

By this result a disagreement between the empirical equation of motion

$$\ddot{\boldsymbol{r}} = -4\pi^2 \frac{\boldsymbol{r}}{r^3}$$

and the hypothetical law of gravitation

$$\ddot{\boldsymbol{r}} = -G(m_0 + m) \frac{\boldsymbol{r}}{r^3}$$

is removed. (Here, m_0 denotes the mass of the Sun, and m that of any planet.) According to the last equation the relative acceleration of a planet depends on its mass, while by the first it is the same for all planets. This leads to the strange result that all planets must have the same mass. This is not, however, an inevitable conclusion. The equation

$$G(m_0 + m) = 4\pi^2$$

has only empirical validity so that it is not necessary that the quantity $[G(m_0 + m) - 4\pi^2]$ should vanish, but only that it should be small compared with the observational errors. Hence

$$| G(m_0 + m) - 4\pi^2 | \ll 4\pi^2$$

for all planets. We now take the mass of the Sun $m_0 = 1$. This choice of the unit of mass is the simplest and most obvious one. Since, now, $m \ll 1$, it follows from the foregoing inequality that

$$\left| \frac{G}{4\pi^2} - 1 \right| \ll 1$$

so that the relative difference of the constant of gravitation from $4\pi^2$ is very small.

The equation for the relative motion of a single planet reads now

$$\ddot{\boldsymbol{r}} = -G(1 + m) \frac{\boldsymbol{r}}{r^3} .$$

On this basis we shall make the following developments. We can, without difficulty, return to the simpler empirical formula by introducing the new time variable

$$\tau = t \sqrt{\left[\frac{G}{4\pi^2}(1 + m)\right]} .$$

(By this we diminish the unit of time by a quantity of the order of magnitude of m, since the interval $t = 0 \ldots t = 1$ corresponds to the interval $\tau = 0 \ldots \tau = \sqrt{[G/(4\pi^2)(1 + m)]}$.) The equation of motion then reads

$$\frac{\mathrm{d}^2\boldsymbol{r}}{\mathrm{d}\tau^2} = -4\pi^2 \frac{\boldsymbol{r}}{r^3} .$$

If

$$\boldsymbol{r} = \boldsymbol{r}(\tau)$$

is any solution of this equation, then

$$\boldsymbol{r} = \boldsymbol{r}\left\{t \sqrt{\left[\frac{G}{4\pi^2}(1 + m)\right]}\right\}$$

is the same solution expressed in terms of the actual time variable t.

2. Systems of ordinary differential equations

1. We come now to the principal problem of dynamics, in its simplest form. We are given the equation for the relative motion of the two mass points

$$\mu \ddot{\boldsymbol{r}} = \boldsymbol{f},$$

where the force $\boldsymbol{f}$ is a known (vector) function of the position $\boldsymbol{r}$, and μ is the sum of the masses. The problem is to evaluate the position vector $\boldsymbol{r}$ as a function $\boldsymbol{r}(t)$ of the time.

Let us introduce the velocity vector $\boldsymbol{v}$ so that we may write the equation of motion as

$$\left.\begin{aligned} \dot{\boldsymbol{r}} &= \boldsymbol{v}, \\ \dot{\boldsymbol{v}} &= \frac{1}{\mu}\boldsymbol{f}(\boldsymbol{r}). \end{aligned}\right\}$$

This is a system of six ordinary differential equations of the first order, which is to be solved for the six unknown functions $\boldsymbol{r}(t)$ and $\boldsymbol{v}(t)$. The principal problem of dynamics is, therefore, the principal problem of the theory of such equations. We shall now formulate the main theorems of this theory.

2. Purely for the sake of simplicity in expression we consider a system of only two equations

$$\left.\begin{aligned} \dot{x} &= f(x, y, t), \\ \dot{y} &= g(x, y, t), \end{aligned}\right\}$$

with two unknown functions $x = \phi(t)$, $y = \psi(t)$. The general case of n equations ($n \geqslant 2$) is handled in an exactly analogous manner, as also the case of only one equation in which we set $g \equiv 0$.

To solve the equations, the method of successive approximations described in I § 5.4 can usefully be applied. For this purpose we write the system of equations in the following manner:

$$\left.\begin{aligned} x &= x_0 + \int_0^t f\,\{x(\tau), y(\tau), \tau\}\,\mathrm{d}\tau, \\ y &= y_0 + \int_0^t g\{x(\tau), y(\tau), \tau\}\,\mathrm{d}\tau, \end{aligned}\right\}$$

where x_0, y_0 denote two arbitrary constants of integration. They are called the initial conditions since, obviously, when $t = 0$

$$x = x_0, \quad y = y_0.$$

The initial conditions can be used as a zero order approximation from which a first approximate solution of the equations may be obtained. Namely

$$\left.\begin{aligned} x_1(t) &= x_0 + \int_0^t f(x_0, y_0, \tau)\,\mathrm{d}\tau, \\ y_1(t) &= y_0 + \int_0^t g(x_0, y_0, \tau)\,\mathrm{d}\tau. \end{aligned}\right\}$$

From these may be obtained a second approximation, and so on. Under certain additional conditions, which in the mechanical applications are practically always fulfilled, the sequence of functions so constructed converges, within a particular and usually narrow time interval, to a solution

$$\left.\begin{aligned} x &= \phi(\mathring{x}, \mathring{y}, t), \\ y &= \psi(\mathring{x}, \mathring{y}, t). \end{aligned}\right\}$$

For the given initial conditions there is no solution besides this one. That is the principal theorem. It may be proved by a method quite similar to that in I § 5.4. This procedure is also very useful for numerical calculations, especially when the functions f and g are (absolutely) small and the relevant time interval is not too large. Under these conditions the functions $\phi(t)$ and $\psi(t)$ will vary very little, so that the right-hand sides will be closely approximated by the constant initial values. Thus, even the functions $x_1(t)$, $y_1(t)$ will give useful approximations for the functions ϕ and ψ, provided $|\, t \,|$ is not too large.

Besides (x_0, y_0), any other initial approximate solution may be used. The better the initial approximation, the quicker will the sequence of successive approximations converge.

3. If the system of equations

$$\left.\begin{aligned} \dot{x} &= f(x, y, t), \\ \dot{y} &= g(x, y, t) \end{aligned}\right\}$$

is such that the right-hand sides may be developed in power series of $(x - x_0)$, $(y - y_0)$ and t, then the general solution may also be obtained in the following manner. We write x and y as power series in t,

$$\left.\begin{aligned} x &= x_0 + a_1 t + a_2 t^2 + \dots, \\ y &= y_0 + a_1 t + a_2 t^2 + \dots, \end{aligned}\right\}$$

and introduce these series into the system of equations. We then develop the right-hand sides in powers of t and equate coefficients on either side. The coefficients a_1, b_1, a_2, b_2, ... are thus uniquely determined. They depend on x_0 and y_0. For given initial values there is at most one solution in the form of a power series. (This is obvious, of course, from the result of subsection 2, but it follows here quite simply.) This quite formally constructed series converges, as may be proved, for sufficiently small $|t|$ and so provides an actual solution.

Sometimes the following method of calculation is more convenient. We expand x and y in Taylor series:

$$\left.\begin{aligned} x &= x_0 + \frac{\dot{x}_0}{1!} t + \frac{\ddot{x}_0}{2!} t^2 + \dots, \\ y &= y_0 + \frac{y_0}{1!} t + \frac{\ddot{y}_0}{2!} t^2 + \dots, \end{aligned}\right\}$$

and evaluate the derivatives $(\ddot{x}, \ddot{y})$, $(\dddot{x}, \dddot{y})$ by successive differentiation of the differential equations:

$$\left.\begin{aligned} \ddot{x} &= f_x \dot{x} + f_y \dot{y} + f_t = f_x \cdot f + f_y \cdot g + f_t, \\ \ddot{y} &= g_x \dot{x} + g_y \dot{y} + g_t = g_x \cdot f + g_y \cdot g + g_t, \end{aligned}\right\}$$

and so on. The suffices x, y, t denote the respective differentiation variables. Finally, on the right-hand side we put $t = 0$, $x = x_0$, $y = y_0$. So we obtain the coefficients of the series, $\dot{x}_0$, $\dot{y}_0$, $\ddot{x}_0$, $\ddot{y}_0$ etc.

4. We shall now apply the procedure just described to the example of the empirical equation of motion of a single planet

$$\ddot{\boldsymbol{r}} = -\frac{4\pi^2}{r^2} \frac{\boldsymbol{r}}{r}.$$

We rewrite the equation

$$\left.\begin{aligned} \dot{\boldsymbol{r}} &= \boldsymbol{v}, \\ \dot{\boldsymbol{v}} &= -\frac{4\pi^2}{r^2} \frac{\boldsymbol{r}}{r}. \end{aligned}\right\}$$

Differentiating and remembering the original equation, we obtain

$$\left.\begin{aligned}\ddot{\boldsymbol{r}} &= -\frac{4\pi^2}{r^2}\cdot\frac{\boldsymbol{r}}{r},\\ \dot{\boldsymbol{v}} &= -\frac{4\pi^2}{r^3}\left(\boldsymbol{v}-\frac{3\dot{r}}{r}\boldsymbol{r}\right),\end{aligned}\right\}$$

$$\left.\begin{aligned}\dddot{\boldsymbol{r}} &= -\frac{4\pi^2}{r^3}\left(\boldsymbol{v}-\frac{3\dot{r}}{r}\boldsymbol{r}\right),\\ \ddot{\boldsymbol{v}} &= .\;\;.\;\;.\;\;.\;\;.\;\;.\;\;.\;\;,\end{aligned}\right\}$$

etc.

We obtain the time derivatives $\dot{r}$, $\ddot{r}$ ... by successive differentiations of the identity

$$r^2 = \boldsymbol{r}^2,$$

which gives

$$\dot{r}r = \dot{\boldsymbol{r}}\boldsymbol{r} = \boldsymbol{v}\boldsymbol{r},$$

$$\ddot{r}r + \dot{r}^2 = \dot{\boldsymbol{v}}\boldsymbol{r} + \boldsymbol{v}\dot{\boldsymbol{r}} = -\frac{4\pi^2}{r} + \boldsymbol{v}^2,$$

etc.

Finally, we take the derivative at the point $t = 0$ (denoted by the suffix 0) and so the power series for $r(t)$ becomes

$$\boldsymbol{r}(t) = \boldsymbol{r}_0 + \frac{1}{1!}\boldsymbol{v}_0 t + \frac{1}{2!}\cdot\frac{-4\pi^2}{r_0^2}\frac{\boldsymbol{r}_0}{r_0}t^2 + \frac{1}{3!}\cdot\frac{-4\pi^2}{r_0^3}\left(\boldsymbol{v}_0 - \frac{3\dot{r}_0}{r_0}\boldsymbol{r}_0\right)t^3 + \dots \qquad (1)$$

This series can be conveniently used for numerical calculation. Moreover it has an important application in theory of the determination of orbits (§ 4.2). We do not obtain a general insight into the form of the motion, except that the orbit must lie in a plane determined by the vectors $\boldsymbol{r}_0$, $\boldsymbol{v}_0$. This much however is evident anyway, and may be easily verified analytically by choosing that plane as the (x, y)-plane and then showing that the function $z(t) \equiv 0$ satisfies both the initial conditions and the equation of motion.

For a single planet we can find the exact orbit. In I § 6.2 we have shown how the elements may be obtained from $\boldsymbol{r}_0$, $\boldsymbol{v}_0$, and in I § 5 how to calculate positions from the elements.

We have obtained the fundamental equation of motion from

the empirically known motions of the planets. We now ask, in reverse: is elliptical motion the only form compatible with the equation of motion, or are there other possible forms? The power series we have just developed can hardly give an answer. In the next section another method is described which is more useful in this respect. It is even possible to survey all the possible forms of motion for an arbitrary law of force of the form $f(r)$. For the Newtonian law of gravitation special simplifications are feasible for the purpose of calculation, so that its consideration differs from that of the general case. We now consider only this special case.

3. The two body problem

1. By §§ 1.4 and 5 the relative motion of two gravitating mass points with the masses m_0 and m_1 obeys the equation

$$\ddot{\boldsymbol{r}} = -\frac{G\mu}{r^2}\frac{\boldsymbol{r}}{r}, \tag{1}$$

where we have abbreviated

$$m_0 + m_1 = \mu. \tag{2}$$

The acceleration $\ddot{\boldsymbol{r}}$ is collinear with the position vector, so we must have

$$\boldsymbol{r} \times \ddot{\boldsymbol{r}} \equiv 0.$$

Now

$$\boldsymbol{r} \times \ddot{\boldsymbol{r}} \equiv \frac{\mathrm{d}}{\mathrm{dt}}(\boldsymbol{r} \times \dot{\boldsymbol{r}}),$$

therefore

$$\boldsymbol{r} \times \dot{\boldsymbol{r}} \equiv 2\boldsymbol{c} \tag{3}$$

where $\boldsymbol{c}$ is a vector which is constant with respect to time.

We can express the fact that the left-hand side is constant with respect to time in another way. Consider any system of ordinary differential equations, for example the system

$$\left.\begin{aligned} \dot{x} &= f(x, y, t),\\ \dot{y} &= g(x, y, t), \end{aligned}\right\}$$

with the general solution

$$\left.\begin{aligned} x &= \phi(a, b, t),\\ y &= \psi(a, b, t),\\ a, b &= \text{const.}, \end{aligned}\right\}$$

then any function $\chi(x, y, t)$ of the *independent* variables x, y, t is called a (first) integral of the system of differential equations if

$$\chi\{\phi(a, b, t), \psi(a, b, t), t\}$$

is constant, with respect to time, for any arbitrary choice of the constants a and b. There is a corresponding definition for the general case of n differential equations.

We can write the equation of motion in the form

$$\left.\begin{aligned} \dot{\boldsymbol{r}} &= \boldsymbol{v}, \\ \dot{\boldsymbol{v}} &= -G\mu \frac{\boldsymbol{r}}{r^3}, \end{aligned}\right\}$$

so that the three components of the vector $\frac{1}{2}\boldsymbol{r} \times \boldsymbol{v}$ are, therefore, first integrals of these six differential equations of motion. They are called the integrals of area, since they represent the areal velocities for the motions of the projections of the two mass points on the coordinate planes.

If we take the scalar product of equation (3) with $\boldsymbol{r}(t)$, the left-hand side becomes the triple scalar product

$$\boldsymbol{r} \,.\, (\boldsymbol{r} \times \boldsymbol{v}) = [\boldsymbol{r}, \boldsymbol{r}, \boldsymbol{v}].$$

This is identical with the determinant of the three vectors, and since two of these are equal the determinant vanishes. Therefore,

$$\boldsymbol{r}(t) \,.\, \boldsymbol{c} \equiv 0$$

—the motion takes place in a plane and the vector of areal velocity, $\boldsymbol{c}$, is normal to this plane. We have seen already, in § 2.4, that the motion is plane.

If we choose the orbital plane as the (ξ, η)-plane, (i.e. as the fundamental coordinate plane—see I § 4.3), we obtain

$$c_\xi = 0, \qquad c_\eta = 0, \qquad c_\zeta = |\,\boldsymbol{c}\,|,$$

and the integrals read

$$\zeta(t) \equiv 0, \qquad \dot{\zeta}(t) \equiv 0,$$

$$\tfrac{1}{2}(\xi\dot{\eta} - \eta\dot{\xi}) \equiv c.$$

We have again obtained *Kepler's second law*, but this time in a form which holds for *each* of the possible motions, and not only for the elliptical motion of the planets. Indeed, still more generally, the law of areas holds for any force $\boldsymbol{f}(r)$ that satisfies the force postulates of § 1.1. It may readily be seen that the derivation still holds for such a general case.

Since $\boldsymbol{c} = \frac{1}{2}\boldsymbol{r} \times \boldsymbol{v}$, we have by I § 6.2

$$\boldsymbol{c} = c\boldsymbol{k} \tag{5}$$

where

$$\boldsymbol{k} = \begin{pmatrix} \sin \Omega \sin \iota \\ -\cos \Omega \sin \iota \\ \cos \iota \end{pmatrix} \tag{6}$$

and c is the magnitude of the areal velocity of the planet in the (ξ, η)-plane. By I § 4.4 and the empirical equations of motion we have, for elliptical motion,

$$c = \pi\sqrt{[(1 - e^2)a]}. \tag{7}$$

In the theoretical equation of motion we have multiplied the time variable by $\sqrt{(G\mu)} / 2\pi$ (§ 1.6), and, therefore, we must also multiply the velocity by this factor. Then we have

$$c = \tfrac{1}{2}\sqrt{[(1 - e^2)a \,.\, G\mu]}. \tag{8}$$

2. We can now obtain a further integral of the equations of motion

$$\ddot{\boldsymbol{r}} = -G\mu \frac{\boldsymbol{r}}{r^3}\,. \tag{1}$$

We take the scalar product of $\dot{\boldsymbol{r}}$ with the second equation of (1). This gives

$$\ddot{\boldsymbol{r}}\dot{\boldsymbol{r}} = \left(-G\mu \frac{\boldsymbol{r}}{r^3}\right) \dot{\boldsymbol{r}}$$

The left-hand side is obviously the time derivative of $\frac{1}{2}\dot{\boldsymbol{r}}^2 = \frac{1}{2}\boldsymbol{v}^2$. The right-hand side is equal to

$$-G\mu\left(\frac{x}{r^3}\,\dot{x} + \frac{y}{r^3}\,\dot{y} + \frac{z}{r^3}\,\dot{z}\right).$$

Inserting the identities

$$\frac{x}{r^3} \equiv -\frac{\partial}{\partial x}\left(\frac{1}{r}\right), \quad \text{etc.,}$$

the right-hand side becomes

$$G\mu\left\{\frac{\partial}{\partial x}\left(\frac{1}{r}\right)\dot{x}+\frac{\partial}{\partial y}\left(\frac{1}{r}\right)\dot{y}+\frac{\partial}{\partial z}\left(\frac{1}{r}\right)\dot{z}\right\}\equiv G\mu\frac{d}{dt}\left(\frac{1}{r}\right).$$

Therefore,

$$\frac{d}{dt}\left(\frac{1}{2}\boldsymbol{v}^2\right)=\frac{d}{dt}\left(\frac{G\mu}{r}\right)$$

or

$$\frac{1}{2}\boldsymbol{v}^2-\frac{G\mu}{r}=h \tag{2}$$

where h denotes some quantity which is independent of the time. The left-hand side, considered as a function of the independent variables $\boldsymbol{v}$ and $\boldsymbol{r}$, is called the energy integral of the equations of motion; the law obtained from it is the energy law. We call the constant h (preliminarily) the energy of the mechanical system.

It can easily be shown that an energy integral exists for any law of force of the form $\boldsymbol{f}(r)$.

3. We come now to the complete solution (the "integration") of the equations of motion

$$\left.\begin{aligned}\ddot{\xi}&=-\frac{G\mu}{r^2}\frac{\xi}{r},\\ \ddot{\eta}&=-\frac{G\mu}{r^2}\frac{\eta}{r}.\end{aligned}\right\} \tag{1}$$

At this point it is convenient to use polar coordinates (r, u) since the force then depends on only one coordinate, the radius vector r. In polar coordinates the laws of energy and of areas read

$$\left.\begin{aligned}\tfrac{1}{2}(\dot{r}^2+r^2\dot{u}^2)-\frac{G\mu}{r}&=h,\\ \tfrac{1}{2}r^2\dot{u}&=c.\end{aligned}\right\} \tag{2}$$

The time variable t does not appear explicitly in these equations, but only as the differential dt. It may, therefore, easily be eliminated. We rewrite the energy law as

$$\frac{1}{2}\left\{\left(\frac{dr}{du}\right)^2+r^2\right\}\left(\frac{du}{dt}\right)^2-\frac{G\mu}{r}=h,$$

and then eliminate from it the angular velocity $\dot{u}$ by means of the law of areas. This gives

$$\frac{1}{2}\left\{\left(\frac{\mathrm{d}r}{\mathrm{d}u}\right)^2 + r^2\right\}\left(\frac{2c}{r^2}\right)^2 - \frac{G\mu}{r} = h. \tag{3}$$

Solving for $\mathrm{d}u/\mathrm{d}r$ we find

$$\frac{\mathrm{d}u}{\mathrm{d}r} = \frac{2c/r^2}{\sqrt{(2h + 2G\mu/r - 4c/r^2)}}. \tag{4}$$

From this we obtain u by integrating with respect to r. The integral may be expressed by means of elementary functions if we use

$$\phi = \frac{2c}{r}$$

as an integration variable. Then we have

$$\begin{aligned} u &= -\int \frac{\mathrm{d}\phi}{\sqrt{[2h + \frac{1}{4}G^2\mu^2/c^2 - (\phi - \frac{1}{2}G\mu/c^2)^2]}} \\ &= \cos^{-1}\left\{\frac{\phi - \frac{1}{2}G\mu/c}{\sqrt{(2h + \frac{1}{4}G^2\mu^2/c^2)}}\right\} + u_0 \end{aligned} \tag{5}$$

where u_0 is a constant of integration. Solving equation (5) for r gives

$$r = \frac{4c^2/(G\mu)}{1 + \sqrt{[1 + 8c^2h/(G^2\mu^2)]}\cos(u - u_0)}. \tag{6}$$

This is the equation of a conic section in polar coordinates (r, u), the origin of which and, therefore, the mass point at rest is in one focus. This equation is usually written

$$r = \frac{p}{1 + e\cos(u - u_0)} \tag{7}$$

where u_0 defines the direction of the major axis, and p denotes the "parameter" of the conic section, that is, the value of r for $w = u - u_0 = \frac{1}{2}\pi$. Comparing this equation with the results of celestial mechanics, we see that

$$e = \left(1 + \frac{8c^2h}{G^2\mu^2}\right)^{1/2}. \tag{8}$$

The orbit is an ellipse, parabola or hyperbola according as e is less than, equal to or greater than one: that is, according as the energy of the system, h, is negative, zero or positive. This is a *generalization of Kepler's first law.*

Parabolic and hyperbolic orbits possibly exist in the Solar System amongst comets and meteors. The eccentricity is always close to unity, so that the hyperbolic orbits deviate only very little from parabolae.

In this section we have tacitly assumed that the areal velocity c is different from zero. We shall briefly consider the case $c = 0$ in section 6. Meanwhile we assume $c > 0$.

4. The law relating the position of a mass point in its orbit to the time can be obtained, exactly as in I § 5.1, by eliminating r from the two equations

$$\left.\begin{aligned} r &= \frac{p}{1 + e\cos w}, \qquad (w = u - u_0) \\ \tfrac{1}{2}r^2\dot{w} &= c. \end{aligned}\right\} \tag{1}$$

This gives, immediately,

$$\tfrac{1}{2}p^2\dot{w} = c(1 + e\cos w)^2, \tag{2}$$

and this differential equation for w can be integrated just as we have already shown in I § 5.1.

Kepler's equation (or its analogues) can also be used. For elliptical motion all the earlier results may be obtained, as in I § 5.2. In particular we have Kepler's equation,

$$E - e\sin E = n(t - T), \tag{3}$$

but with the difference that the mean motion,

$$n = \frac{2\pi}{P} \qquad (P = \text{period}), \tag{4}$$

is now given by

$$n = \frac{\sqrt{(G\mu)}}{a^{3/2}}. \tag{5}$$

(The factor μ comes from the new expression for the areal velocity

$$c = \tfrac{1}{2}\sqrt{[G\mu a(1 - e^2)]} \tag{6}$$

of section 2.) Equating the right-hand sides of equations (4) and

(5), we obtain a *generalized and more rigorous form of Kepler's third law*, namely

$$a^3 = \frac{G}{4\pi^2}\,\mu P^2 . \tag{7}$$

5. Hyperbolic orbits may be treated in a similar manner. Parabolic orbits, however, are particularly simple, and, on account of their relevance to comets, we shall consider them in some detail. Since $e = 1$ and

$$1 + \cos w = 2\cos^2\frac{w}{2},$$

equation 4(2)

$$\tfrac{1}{2}p^2\dot{w} = c(1 + e\cos w)^2$$

becomes

$$\frac{p^2}{8c}\,\dot{w} = \cos^4\frac{w}{2} \tag{1}$$

or

$$\frac{dw}{\cos^4 w/2} = \frac{8c}{p^2}\,dt . \tag{2}$$

Both sides may be integrated by means of elementary functions, to give

$$\tan\frac{w}{2} + \frac{1}{3}\tan^3\frac{w}{2} = \frac{4c}{p^2}(t - T), \tag{3}$$

where T again denotes the time of perihelion. (This equation can easily be shown to be correct by differentiating it.)

The coefficient $4c/p^2$ can be further simplified. For an ellipse,

$$p = a(1 - e^2)$$

by 3(7) and

$$c = \tfrac{1}{2}\sqrt{[G\mu a(1 - e^2)]}$$

by 4(6). Therefore

$$\frac{4c}{p^2} = \frac{2\sqrt{(G\mu)}}{p^{3/2}} . \tag{4}$$

Since the parabola is a limiting case of the ellipse, this equation still holds ($e \to 1$). Also, for the parabola,

$$p = 2q,$$

where q is the "perihelion distance". So, finally, we have for the true anomaly w the relation

$$\tan\frac{w}{2}+\frac{1}{3}\tan^3\frac{w}{2}=\frac{1}{\sqrt{2}}\frac{\sqrt{(G\mu)}}{q^{3/2}}(t-T)\,. \tag{5}$$

The representation of the Cartesian coordinates as functions of the time and of the elements follows exactly the method of I § 5, and need not, therefore, be repeated here.

6. The relation between r and t can also be derived in a similiar way to the relation between r and u in section 3. This time du is eliminated from the laws of energy and of areas, instead of dt. This gives

$$\dot{r}^2=2h+2\frac{G\mu}{r}-\frac{4c^2}{r^2} \tag{1}$$

or

$$\frac{\mathrm{d}t}{\mathrm{d}r}=\frac{1}{\sqrt{(2h+2G\mu/r-4c^2/r^2)}}\,. \tag{2}$$

This equation is also valid for the case $c = 0$, hitherto excluded. From the law of areas

$$\tfrac{1}{2}r^2\dot{u}=c \tag{3}$$

we see that if, and only if, $c = 0$ the motion is on a straight line passing through the point at rest. Then and only then, is a collision possible. The form of the motion, apart from the collision, is clear if the rectilinear motion is understood as a limiting case of motion in a conic section. (We also need the fictitious supposition that the mass points are really points, which can penetrate each other without resistance and without any kind of alteration.)

When $c = 0$ and r is very small, dt/dr will be asymptotically proportional to $\sqrt{r}$. If t_0 is the time of the collision, then $t - t_0$ is asymptotically proportional to $r^{3/2}$, and r to $(t - t_0)^{2/3}$. The function $r(t)$, therefore, has an algebraic singularity at the point t_0.

Similarly, for arbitrary c and parabolic motion ($h = 0$), we obtain from the equation for dt/dr

$$\frac{\mathrm{d}t}{\mathrm{d}r}\simeq\sqrt{\frac{r}{2G\mu}}\qquad\text{for } r\to\infty,$$

($\simeq$ denotes asymptotic equality)

so that

$$\dot{r} \simeq \left(\frac{2G\mu}{r}\right)^{1/2} \qquad \text{for } r \to \infty$$

and

$$|t| \simeq \frac{2}{3}\frac{r^{3/2}}{\sqrt{(2G\mu)}} \qquad \text{for } r \to \infty,$$

$$r \simeq [\tfrac{3}{2}\sqrt{(2G\mu)}\,|t|]^{2/3} \quad \text{for } |t| \to \infty.$$

Similarly, for hyperbolic motion ($h > 0$),

$$\left.\begin{aligned} \dot{r} &\simeq \sqrt{(2h)}, \\ r &\simeq \sqrt{(2h)}\,|t|, \end{aligned}\right\} \qquad \text{for } |t| \to \infty.$$

For large $|t|$, therefore, the motion is nearly both rectilinear and uniform.

7. We wish to derive a further important relation for elliptical motion. In the purely geometrical relation

$$p = a(1 - e^2) \tag{1}$$

(see I § 5.1 and II § 2), we substitute the dynamical quantities as they are obtained from comparison of the polar equations 3(6) and (7) of the orbital ellipse, namely,

$$r = \frac{p}{1 + e \cos w} \tag{2}$$

and

$$r = \frac{4c^2/(G\mu)}{1 + \sqrt{[1 + 8c^2h/(G^2\mu^2)]}\cos w}. \tag{3}$$

These give

$$p = \frac{4c^2}{G\mu} \tag{4}$$

and

$$e^2 = 1 + 8\frac{c^2h}{G^2\mu^2}, \tag{5}$$

and so the geometrical relation yields, after a few steps,

$$a = -\frac{G\mu}{2h} \tag{6}$$

—the semi-major axis a is, for a given total mass μ, determined by the energy, h, alone.

By the generalization of Kepler's third law,

$$a^3 = \frac{G}{4\pi^2}\,\mu P^2, \tag{7}$$

the period P is also, for a given total mass μ, a function of the energy h alone. This is a special case of a more general theorem of mechanics which, however, we shall not consider any further here.

8. Kepler's third law in its new form 7(7) can now be used to determine the exact value of the constant of gravitation G. We apply the law to the Earth, the mass of which, m_{E}, is very much less than unity. Since, for the Earth,

$$a = 1, \quad P = 1, \quad \mu = 1 + m_{\mathrm{E}}$$

we have, by 7(7),

$$G = \frac{4\pi^2}{1 + m_{\mathrm{E}}}. \tag{1}$$

We still have to determine the mass of the Earth, m_{E}, however. To do this we observe the motion of the Moon around the Earth. At this point we introduce the assumption that the satellites also move around their respective planets as if they were subjected to the Newtonian forces as exerted by mass points at the centres of the planets. This assumption holds only approximately. The moons are so near to the planets, that the assumption that the planets are mass points must be qualified. Further, superimposed on the attraction of the primary planet is the weak, but still noticeable, gravitational attraction exerted by the neighbouring planets and, above all, the Sun.

The sidereal period of the Moon can be obtained from the observed "synodic" period similarly as for the planets themselves. The semi-major axis must be obtained from the determination of the orbit. This can be done, in principle, by the method of I § 6.4, the mass being treated as a seventh element, to be improved equally with the others. First approximations for the unknowns can be obtained from the observations if the orbit is assumed to be circular which, in fact, it almost is.

The mass of a moon is small compared with that of its planet. Thus, for planets with several moons, the different moons must yield the same value for the total mass.

In the following table we present for each planet the number of moons, the mass, the diameter and the mean density. For comparison we include the Sun at the end. The masses of Mercury and Venus, which possess no satellites, must be estimated from the perturbations and so are not very certain. (Also the diameters of the distant planets and therefore their mean densities are not very well known.)

Planet	No. of moons	Mass ($\odot = 1$)	Mass (Earth = 1)	Diameter (Earth = 1)	Mean density (Earth = 1)
Mercury	—	$1{\cdot}6 \times 10^{-7}$	$5{\cdot}3 \times 10^{-2}$	0·40	~1
Venus	—	$2{\cdot}4 \times 10^{-6}$	$8{\cdot}15 \times 10^{-1}$	0·99	~0·9
Earth	1	$3{\cdot}0 \times 10^{-6}$	1	1	1
Mars	2	$3{\cdot}2 \times 10^{-7}$	0·11	0·54	0·70
Jupiter	12	$9{\cdot}5 \times 10^{-4}$	318	11	0·24
Saturn	9 or 10, ring	$2{\cdot}9 \times 10^{-4}$	95	9	0·12
Uranus	5	$4{\cdot}4 \times 10^{-5}$	15	4·2	0·20
Neptune	2	$5{\cdot}2 \times 10^{-5}$	17	3·9	0·29
Pluto	?				
Sun	—	1	$3{\cdot}3 \times 10^{5}$	109	0·26

These data demonstrate a rough law: the number of moons, the mass and the diameter increase (with the partial exception of Mars) until Jupiter, and then decrease again, while the mean densities of Jupiter and of the more distant planets are of an order of magnitude less than that of the Earth. From the value for the mass of the Earth we finally see that the constant of gravitation

$$G = \frac{4\pi^2}{1 + m_E} \tag{2}$$

differs from $4\pi^2$ by only about one part in 300,000.

4. The determination of orbits

1. In I § 6.4 we supposed that approximate values of the elements of the orbit of a planet were already available, and then we showed how to improve these values. We did not consider the problem of finding these first approximations themselves.

Of course we could have done so, since the problem is of a purely kinematical nature. We have left it until now, however, since the use of dynamical results simplifies the solution.

Let the directions, $\boldsymbol{u}$, of the planet in question be observed at the instants t_i, $i = 1, 2, 3$. If $\boldsymbol{q}$ denotes the position vector of the Earth, and $\boldsymbol{r}$ that of the planet, and if ϱ is the distance of the planet from the Earth, $q = |\boldsymbol{q}|$ and $r = |\boldsymbol{r}|$, then

$$\boldsymbol{q} + \varrho \boldsymbol{u} = \boldsymbol{r} \tag{1}$$

at all times. (A correction must be made for the aberration time, cf. I § 6.8.) The basic idea of one method is to differentiate this equation twice with respect to the time variable t, and then to express the resulting acceleration in terms of the force, by means of the equation of motion. First we obtain, by differentiation,

$$\ddot{\varrho}\boldsymbol{u} + 2\dot{\varrho}\dot{\boldsymbol{u}} + \varrho\ddot{\boldsymbol{u}} = \ddot{\boldsymbol{r}} - \ddot{\boldsymbol{q}} \tag{2}$$

which gives, using the equation of motion,

$$\ddot{\varrho}\boldsymbol{u} + 2\dot{\varrho}\dot{\boldsymbol{u}} + \varrho\ddot{\boldsymbol{u}} = -4\pi^2\left(\frac{\boldsymbol{r}}{r^3} - \frac{\boldsymbol{q}}{q^3}\right). \tag{3}$$

(We neglect the masses of the planets compared with that of the Sun.) Now, for the middle instant t_2, we see that not only the direction of the planet, $\boldsymbol{u}$, is known, but also (approximately) its angular velocity $\dot{\boldsymbol{u}}$ and angular acceleration $\ddot{\boldsymbol{u}}$. We can obtain these two vectors from the three known values of the function $\boldsymbol{u}(t)$ by numerical differentiation. Our aim is to evaluate the distance ϱ and the radial velocity $\dot{\varrho}$ at the time t_2. If we know these two, then we also know the position vector

$$\boldsymbol{r} = \boldsymbol{q} + \varrho\boldsymbol{u}$$

and the velocity vector

$$\dot{\boldsymbol{r}} = \dot{\boldsymbol{q}} + \dot{\varrho}\boldsymbol{u} + \varrho\dot{\boldsymbol{u}} \tag{4}$$

at the time t_2. (The motion of the Earth, and therefore the vectors $\boldsymbol{q}$ and $\dot{\boldsymbol{q}}$ are assumed to be known.) From $\boldsymbol{r}_2$ and $\dot{\boldsymbol{r}}_2$ we can easily calculate the orbit of the planet, either directly by the method of § 2.4, or through the elements, which may be evaluated by the method of I § 6.2.

We now obtain one equation each for ϱ and $\dot{\varrho}$ by taking the scalar product of equation (3) first with $(\boldsymbol{u} \times \dot{\boldsymbol{u}})$ and then with $(\boldsymbol{u} \times \ddot{\boldsymbol{u}})$. Thus,

$$[\boldsymbol{u}, \dot{\boldsymbol{u}}, \ddot{\boldsymbol{u}}]\varrho = 4\pi^2 \left\{\frac{1}{q^3}[\boldsymbol{u}, \dot{\boldsymbol{u}}, \boldsymbol{q}] - \frac{1}{r^3}[\boldsymbol{u}, \dot{\boldsymbol{u}}, \boldsymbol{r}]\right\} \tag{5}$$

and

$$2[\boldsymbol{u}, \ddot{\boldsymbol{u}}, \dot{\boldsymbol{u}}]\dot{\varrho} = 4\pi^2 \left\{\frac{1}{q^3}[\boldsymbol{u}, \ddot{\boldsymbol{u}}, \boldsymbol{q}] - \frac{1}{r^3}[\boldsymbol{u}, \ddot{\boldsymbol{u}}, \boldsymbol{r}]\right\}. \tag{6}$$

The square brackets $[\boldsymbol{a}, \boldsymbol{b}, \boldsymbol{c}]$ denote, as in § 3.1, the volume enclosed by the (orientated) parallelepiped defined by the three arbitrary vectors $\boldsymbol{a}, \boldsymbol{b}, \boldsymbol{c}$.

The formulae for ϱ and $\dot{\varrho}$ can still be simplified since

$$\begin{aligned}[\boldsymbol{u}, \dot{\boldsymbol{u}}, \boldsymbol{r}] &= [\boldsymbol{u}, \dot{\boldsymbol{u}}, \boldsymbol{q} + \varrho\boldsymbol{u}] \\ &= [\boldsymbol{u}, \dot{\boldsymbol{u}}, \boldsymbol{q}] + \varrho[\boldsymbol{u}, \dot{\boldsymbol{u}}, \boldsymbol{u}] \\ &= [\boldsymbol{u}, \dot{\boldsymbol{u}}, \boldsymbol{q}]\end{aligned}$$

and, similarly

$$[\boldsymbol{u}, \ddot{\boldsymbol{u}}, \boldsymbol{r}] = [\boldsymbol{u}, \ddot{\boldsymbol{u}}, \boldsymbol{q}].$$

So, finally, the formulae for ϱ and $\dot{\varrho}$ read:

$$\left.\begin{aligned}[\boldsymbol{u}, \dot{\boldsymbol{u}}, \ddot{\boldsymbol{u}}]\varrho &= 4\pi^2 \left(\frac{1}{q^3} - \frac{1}{r^3}\right)[\boldsymbol{u}, \dot{\boldsymbol{u}}, \boldsymbol{q}], \\ [\boldsymbol{u}, \dot{\boldsymbol{u}}, \ddot{\boldsymbol{u}}]\dot{\varrho} &= -2\pi^2 \left(\frac{1}{q^3} - \frac{1}{r^3}\right)[\boldsymbol{u}, \ddot{\boldsymbol{u}}, \boldsymbol{q}].\end{aligned}\right\} \tag{7}$$

We have still to eliminate the unknown r from the right-hand sides. To do this we take the scalar square of the fundamental equation (1), which gives

$$r^2 = q^2 + 2\varrho\boldsymbol{q}\boldsymbol{u} + \varrho^2. \tag{8}$$

We introduce this expression into the first equation (7) which now contains only one unknown, ϱ. It is easily put in the form

$$\left(\frac{1}{4\pi^2}\frac{[\boldsymbol{u}, \dot{\boldsymbol{u}}, \ddot{\boldsymbol{u}}]}{[\boldsymbol{u}, \dot{\boldsymbol{u}}, \boldsymbol{q}]}\varrho - \frac{1}{q^3}\right)^2 \cdot (q^2 + 2\boldsymbol{q}\boldsymbol{u} + \varrho^2)^3 = 1 \tag{9}$$

which is an algebraic equation of the eighth degree and can be solved, for example, by the rule of false position. If the equation has more than one positive root for ϱ, then a fourth observation is needed to decide which is correct. The calculation of $\dot{\varrho}$, and all the subsequent steps we have already described.

We have presented only the basic ideas of this method, which was invented by Laplace. In its application there are many small improvements that can be made. Its practical usefulness, however, is disputed. That it can never be very exact is obvious, since the empirical function $\boldsymbol{u}(t)$ is twice differentiated and, as a result, (as we have already made clear in § 2.3) considerable uncertainty is involved. In fact, Laplace's method does not solve the problem as stated, namely to determine the orbit from the three directions $\boldsymbol{u}_1, \boldsymbol{u}_2, \boldsymbol{u}_3$; instead, it solves a different problem: assuming the vectors $\boldsymbol{u}, \dot{\boldsymbol{u}}, \ddot{\boldsymbol{u}}$ to be known at some particular instant, determine the orbit. Gauss devised a method, which many workers prefer, that solves the original problem. It is presented in what follows.

2. We start from the condition that the three unknown positions of a planet, $\boldsymbol{r}_1, \boldsymbol{r}_2, \boldsymbol{r}_3$, must belong to a plane which passes through the Sun. We write this condition in the form

$$\begin{vmatrix} x_1 & x_2 & x_3 \\ y_1 & y_2 & y_3 \\ z_1 & z_2 & z_3 \end{vmatrix} = 0 \qquad (1)$$

or, if we expand the determinant by its first row,

$$x_1 \begin{vmatrix} y_2 & y_3 \\ z_2 & z_3 \end{vmatrix} + x_2 \begin{vmatrix} y_3 & y_1 \\ z_3 & z_1 \end{vmatrix} + x_3 \begin{vmatrix} y_1 & y_2 \\ z_1 & z_2 \end{vmatrix} = 0. \qquad (2)$$

Two analogous equations can be obtained by cyclic interchange of the letters x, y, z.

The coefficients of equation (2) are proportional to the areas contained by the projections of the triangles $(0, \boldsymbol{r}_2, \boldsymbol{r}_3)$, $(0, \boldsymbol{r}_3, \boldsymbol{r}_1)$, $(0, \boldsymbol{r}_1, \boldsymbol{r}_2)$ on the (y, z)-plane. They are, therefore, also proportional to the areas F_1, F_2, F_3 of these triangles themselves, since they all lie in the same plane—the orbital plane. We can, therefore, write

$$F_1 x_1 + F_2 x_2 + F_3 x_3 = 0.$$

Corresponding equations hold for y and z. We combine all three equations in the one vector equation

$$F_1 \boldsymbol{r}_1 + F_2 \boldsymbol{r}_2 + F_3 \boldsymbol{r}_3 = 0. \qquad (3)$$

In equation (3) we substitute for the position vector $\boldsymbol{r}_k$ the expression

$$\boldsymbol{r}_k = \boldsymbol{q}_k + \varrho_k \boldsymbol{u}_k, \qquad k = 1, 2, 3 \qquad (4)$$

and so we obtain

$$F_1\varrho_1\boldsymbol{u}_1 + F_2\varrho_2\boldsymbol{u}_2 + F_3\varrho_3\boldsymbol{u}_3 = -(F_1\boldsymbol{q}_1 + F_2\boldsymbol{q}_2 + F_3\boldsymbol{q}_3). \quad (5)$$

This sytem of equations can be solved for ϱ_k if we take its scalar product with $(\boldsymbol{u}_1 \times \boldsymbol{u}_2)$ etc. Thus,

$$[\boldsymbol{u}_1, \boldsymbol{u}_2, \boldsymbol{u}_3]F_3\varrho_3 = -F_1[\boldsymbol{q}_1, \boldsymbol{u}_1, \boldsymbol{u}_2] - F_2[\boldsymbol{q}_2, \boldsymbol{u}_1, \boldsymbol{u}_2] - - F_3[\boldsymbol{q}_3, \boldsymbol{u}_1, \boldsymbol{u}_2], \quad (6)$$

and two analogous equations hold, which may be obtained by cyclic interchange of the indices 1, 2, 3.

Now we still have to express the area F_k in terms of the time. For example, F_3 is proportional to the magnitude of the vector product $\boldsymbol{r}_1 \times \boldsymbol{r}_2$. Now by II § 2.4(1) we have

$$\boldsymbol{r}_i = \boldsymbol{r}_0 + \boldsymbol{v}_0 t_i - \frac{2\pi^2}{r_0^3}\boldsymbol{r}_0 t_i^2 - \frac{2}{3}\frac{\pi^2}{r_0^3}\left(\boldsymbol{v}_0 - \frac{3\dot{r}_0}{r_0}\boldsymbol{r}_0\right)t_i^3 \quad (7)$$

excluding terms of the fourth and higher orders in t_i. With the same degree of approximation, we obtain by multiplication

$$\boldsymbol{r}_1 \times \boldsymbol{r}_2 = (\boldsymbol{r}_0 \times \boldsymbol{v}_0)\left(\tau_3 - \frac{2}{3}\frac{\pi^2}{r_0^3}\tau_3^3\right), \text{ etc.} \quad (8)$$

where

$$\tau_3 = t_2 - t_1, \text{ etc.} \quad (9)$$

Therefore F_k is proportional to

$$\left(\tau_k - \frac{2}{3}\frac{\pi^2}{r_0^3}\tau_k^3\right),$$

so that the equation for ϱ_k reads

$$[\boldsymbol{u}_1, \boldsymbol{u}_2, \boldsymbol{u}_3]\left(\tau_2 - \frac{2}{3}\frac{\pi^2}{r_0^3}\tau_2^3\right)\varrho_2$$

$$= -\sum_{j=1}^{3}\left(\tau_j - \frac{2}{3}\frac{\pi^2}{r_0^3}\tau_j^3\right)[\boldsymbol{q}_j, \boldsymbol{u}_3, \boldsymbol{u}_1], \text{ etc.} \quad (10)$$

or

$$[\boldsymbol{u}_1, \boldsymbol{u}_2, \boldsymbol{u}_3]\tau_2\varrho_2 + \sum_{j=1}^{3}[\boldsymbol{q}_j, \boldsymbol{u}_3, \boldsymbol{u}_1]\tau_j$$

$$= \frac{2}{3}\frac{\pi^2}{r_0^3}\left\{[\boldsymbol{u}_1, \boldsymbol{u}_2, \boldsymbol{u}_3]\tau_2^3\varrho_2 + \sum_{j=1}^{3}[\boldsymbol{q}_j, \boldsymbol{u}_3, \boldsymbol{u}_1]\tau_j^3\right\}, \text{ etc.} \quad (11)$$

Now we choose as our zero of time the instant t_2 of the middle observation, so that $\boldsymbol{r}_0 = \boldsymbol{r}_2$. Taking the scalar square of

$$\boldsymbol{r}_2 = \boldsymbol{q}_2 + \varrho_2 \boldsymbol{u}_2$$

we have

$$r_2^2 = q_2^2 + 2\varrho_2 \boldsymbol{q}_2 \boldsymbol{u}_2 + \varrho_2^2.$$

From this equation and that for ϱ_2 we eliminate r_2 and thereby obtain for ϱ_2 an algebraic equation of the eighth degree, as in section 1, namely the equation:

$$\left\{[\boldsymbol{u}_1, \boldsymbol{u}_2, \boldsymbol{u}_3]\tau_2\varrho_2 + \sum_{j=1}^{3} [\boldsymbol{q}_j, \boldsymbol{u}_3, \boldsymbol{u}_1]\tau_j\right\}^2 (q_2^2 + 2\boldsymbol{q}_2\boldsymbol{u}_2\varrho_2 + \varrho_2^2)^3$$
$$= \left(\frac{2}{3}\pi^2\right)^2 \left\{[\boldsymbol{u}_1, \boldsymbol{u}_2, \boldsymbol{u}_3]\tau_2^3\varrho_2 + \sum_{j=1}^{3} [\boldsymbol{q}_j, \boldsymbol{u}_3, \boldsymbol{u}_1]\tau_j^3\right\}^2, \quad (12)$$

which can be solved, for instance, by the rule of false position. If τ_2 is sufficiently small, then the pair of equations for ϱ_2 and r_2 can be solved by means of successive approximations, starting with the value $r_2 = \infty$, or, better, from an empirically plausible initial value of r_2. From this an approximate value of ϱ_2 is calculated, and from this a new approximation to r_2 and ϱ_2 and so on.

If r_2 and ϱ_2 are known then the distances ϱ_1 and ϱ_3 can readily be determined from the corresponding equations (11) and, finally, with their help we obtain the three position vectors $\boldsymbol{r}_1$, $\boldsymbol{r}_2$ and $\boldsymbol{r}_3$. The elements of the orbit can be evaluated by the procedure of I § 6.1.

On account of the quantities neglected, only approximate values can be obtained. These can, in principle, be improved by the method of I § 6.5. We shall not go in to the practical details.

5. Summary

Kepler's first law suggests the idea that the Sun is the seat of a motive force. We describe the idea of force formally by means of a series of postulates which are drawn from rough, elementary notions. For example, the force between two mass points (we do not consider any other forces at all) acts on the line joining the two points, and its magnitude depends only on their distance from each other. For Kepler motion, the vector of acceleration satisfies these postulates. This leads to preliminary laws of gravitation and of motion. If the law of gravitation is to hold quite generally between all mass points, then a correction is needed so that the force (or acceleration) depends also on the respective pair of points. The (questionable) idea of an absolute space provides a clue for this correction. The results, Newton's laws of gravitation and of motion, are working hypotheses which must be tested by their fruitfulness. The masses of the planets are very small compared with that of the Sun (§ 1).

The principal problem of theoretical mechanics, given the force to evaluate the motion from the equations of motion, is a problem in the theory of ordinary differential equations. The principal theorem of this theory is: under particular assumptions which we can always take as justified, a system of equations possesses exactly one solution for each set of initial conditions. This solution can always be found, from an arbitrary approximation, either by means of successive approximations or, for analytical equations, by expanding the right-hand sides in power series (§ 2).

In the two body problem four integrals may readily be obtained—the three area integrals and the energy integral. From these we can obtain the solution of the equation of motion in terms of elementary expressions—although only in a parametric representation. Besides the Kepler ellipse, parabolic and hyperbolic orbits are possible. Which case actually occurs is decided by the sign of the energy constant. It is assumed here that the areal velocity is different from zero; otherwise all three forms of motion degenerate to rectilinear motion (§ 3).

Finally, we have presented, in their chief characteristics. both Laplace's and Gauss's methods of solving the problem of determining the first approximations for the orbital elements of a planet or comet from three observations of its direction. Laplace substituted for the original problem this other: to determine the elements from the known values of the direction, angular velocity and angular acceleration, at any one instant. In practice this method must always be inexact, since the angular velocity and the angular acceleration can themselves be only poorly determined from the observed directions. Gauss's method avoids this disadvantage. In this method the position vector of the planet is developed in power series of the time which is cut off after the third order term. Consequently only approximate values are obtained, which need further improvement. (§ 4.)

Exercises

1. What is the expression for Newton's law of gravitation when the mass, length and time are measured in arbitrary units (§ 1)?

2. Derive the law of gravitation from the empirical third law of Kepler, assuming that the planetary orbits are circular. Newton's general law of motion may be assumed. At what points in the argument is induction necessary (§ 1)?

3. What motion would a mass point describe, in absolute space, if no forces at all are acting on it? This result is called the "Law of Inertia."

4. We shall call a system of material points "massless" if there are no forces acting between the individual points. How do the points move relatively to each other? These results may also be formulated as an "Inertial Law". Which formulation is to be preferred, this or that of question 3 (§ 1)?

5. Integrate the differential equations

$$\ddot{x} = x \quad \text{and} \quad \ddot{x} = -x$$

(a) by the method of successive approximations,
(b) by expansion in a power series.
Estimate the error made by breaking off the calculation after a finite number of steps. For what values of t does the power series solution converge? What functions are represented by them (§ 2)?

6. What relation do the solutions of the differential equations

$$\ddot{x} = \alpha x, \qquad \alpha = \text{const.}$$

have to the solutions of the equations in the last question?

7. Transform the energy and (plane) area integrals from Cartesian to polar coordinates (§ 3).

8. For hyperbolic motion the angle between the asymptotes is given by

$$2 \tan^{-1}\left(2\sqrt{2}\,\frac{c\sqrt{h}}{\mu G}\right).$$

Prove and discuss this formula (§ 3.4).

9. From the formula of the last problem we derive the following. Consider light to be made up of small material particles. Then, a light ray from a fixed star (i.e. from infinity) which grazes the limb of the Sun will be deviated through the angle

$$\frac{2\,G\mu}{R\gamma^2},$$

where R is the radius of the Sun, and γ the velocity of light. Determine the numerical value of the deviation. ($G = 4\pi^2$, $\mu = 1$, $R = 7 \times 10^{10}$ cm,

$\gamma = 3 \times 10^{10}$ cm/sec. For the interchange of units see I § 3.3.) Relativity theory predicts exactly double this value—in rough agreement with the observations.

10. According to Newton's assumption all mass points attract each other, so that the assumptions of question 4 are never actually realized. In addition to the earlier questions, formulate a new inertial law by means of the limiting process $\mu \to 0$.

11. Solve the hyperbolic case $h > 0$, $c > 0$ of the two body problem by exact analogy with the elliptical case (§ 3).

12. Solve the differential equation for r in § 3.6 in an analogous manner to that for u in § 3.4.

13. Discuss the three types of rectilinear motion, (elliptic, parabolic and hyperbolic) (§ 3).

14. What is the value of the energy constant for circular motion? What is the magnitude of its circular velocity (§ 3)?

15. The escape velocity at the point $\mathbf{r}$ is the velocity that a moving particle must have if its energy $h = 0$. What is the ratio of the escape velocity to the circular velocity at the same point (§ 3)?

16. What is the solution of the two body problem if a repulsive Newtonian force is considered (§ 3)?

17. Derive the formula by which the sidereal period of the moon can be obtained from its synodic period (§ 3, I § 1).

18. Two stars attracting each other according to Newton's law of gravitation revolve around each other in Kepler ellipses. Their velocities are projected onto a fixed direction in space. These velocity components can be developed in Fourier series. Show how the orbital elements and the masses of such a "spectroscopic binary" may be determined from the first terms of these Fourier series (as far as possible).

19. Can the accuracy of Laplace's method of determining orbits be increased by making the three observations as close together as possible (§ 4)?

20. How must the methods of Gauss and Laplace be modified for the determination of the orbit of the Moon (§ 4)?

21. Discuss the error in Gauss's method in the case of a circular orbit (§ 4).

22. How are the methods of the determination of orbits to be modified for parabolic and hyperbolic orbits (§ 4 and I § 6)?

NOTES

§ **1.1.** Every student should have looked at Newton's *Principia.* The following book clarifies the fundamental problems of mechanics, and is also thought-provoking:

[11] Mach, E., *Die Mechanik in ihrer Entwicklung*, Leipzig, 1889.

A different view is expressed in [2]. An excellent survey over all related problems is given in

[12] Voss, A., Die Prinzipien der rationellen Mechanik. (*Encyclopädie der mathematischen Wissenschaften* IV 1.)

§ **1.3.** On the differences between "exact" and "approximate" mathematics, cf.

[13] KLEIN, F., *Elementarmathematik vom höheren Standpunkt aus*, Berlin, 1924.

On the anomaly of Mercury's motion see any book on relativity, e.g.

[14] EDDINGTON, A. S., *Space, Time and Gravitation*, Cambridge, 1920.

[15] BERGMANN, P. G., *Introduction to the Theory of Relativity*, New York, 1942.

[16] LAUE, M. v., *Die Relativitätstheorie*, Braunschweig, 1955.

§ **1.4.** A more detailed description of the foundations of celestial mechanics is given in [2].

§ **2.4.** For the two body problem with any law of force, cf.

[17] CONTOPOULOS, G., *Z. f. Astrophysik* **35,** 67, (1954), or

[18] KURTH, R., *Astr. Nachr.* **282,** 241, (1956).

§ **3.1–3.** For a general law of force, the derivations of all equations up to § 3.3(4) for du/dr are quite similar. The integration that immediately follows this equation, however, uses the explicit form of the law of force and cannot be generalized.

§ **3.7.** On the theorem mentioned at the end of the section, cf. Wintner [7].

§ **3.8.** Data are taken from Allen's book [5].

CHAPTER III

THE DYNAMICS OF THE PLANETARY SYSTEM

1. The equations of motion and their integrals

1. We now take into consideration the influence of the planets upon each other. We start from the general law of motion of II § 1.4:

$$m_i \ddot{\boldsymbol{r}}_i = \boldsymbol{f}_i, \qquad i = 1, 2, \ldots, n. \tag{1}$$

Here m_i denotes the mass, $\boldsymbol{r}_i$ the position vector, and $\ddot{\boldsymbol{r}}_i$ the acceleration vector of the ith mass point, while $\boldsymbol{f}_i$ is the total force acting upon it. We shall consider the system as mechanically closed—that is to say that apart from the forces of interaction (here, apart from the gravitational interactions) no other force influences the system.

First of all we must write down the explicit expression for the force $\boldsymbol{f}_i$. How are the different forces acting simultaneously on a mass point to be added together? As the answer to this question we give the following postulate which demands the undisturbed superposition of the different forces.

Fifth and last force postulate: The total force exerted on a mass point 0 by a system of mass points 1, 2, ..., n is equal to the vector sum $\sum_{i=1}^{n} \boldsymbol{f}_{i0}$ of the single forces.

This postulate is compatible with everyday experience, although it seems actually plausible only for collinear forces. On account of its simplicity we take it as a working hypothesis, and write, experimentally,

$$\boldsymbol{f}_i = \sum_{j=1}^{n} \boldsymbol{f}_{ji},$$

where

$$\boldsymbol{f}_{ji} = G m_i m_j \frac{\boldsymbol{r}_j - \boldsymbol{r}_i}{|\boldsymbol{r}_j - \boldsymbol{r}_i|^3}, \qquad j \neq i,$$

signifies the force exerted on the point i by the point j. For the sake of formal simplicity we have just introduced a force

$$\boldsymbol{f}_{ii} \equiv 0, \qquad i = 1, 2, \ldots, n.$$

The way to test Newtonian celestial mechanics is now clear: the equations of motion

$$m_i \ddot{\boldsymbol{r}}_i = \sum_{j=1}^{n} \boldsymbol{f}_{ji}, \qquad i = 1, 2, \ldots, n, \tag{2}$$

have to be integrated, and the solutions $\boldsymbol{r}_i(t)$ compared with the observations. For us, therefore, the problem is to solve the foregoing system of differential equations.

2. We introduce now the velocity vector

$$\boldsymbol{v}_i = \dot{\boldsymbol{r}}_i.$$

Then, the system of the equations of motion

$$\left.\begin{aligned} \dot{\boldsymbol{r}}_i &= \boldsymbol{v}_i \\ \dot{\boldsymbol{v}}_i &= \frac{1}{m_i}\boldsymbol{f}_i \end{aligned}\right\} i = 1, 2, \ldots, n,$$

may be solved by the method of II § 2, at least in principle. We have already had an example of this in the two body problem, which can be solved numerically in this way to an arbitrary degree of accuracy, although hardly any insight into the nature of the solution can thus be obtained. We will, therefore, as in that example, try to take account of the particular nature of the dynamical problem. This will make the calculation easier and lead us to definite assertions.

First, however, a comment must be made. Among the assumptions upon which the fundamental theorems and methods of II § 2 rest, there is the assumption that the right-hand sides of the differential equations remain bounded. Now if, in the course of time two or more mass points collide (simultaneously), then this assumption is certainly no longer correct. To obtain the solution during the collision of two particles, say particles 1 and 2, we suppose that the motion shortly before and shortly after the collision is essentially determined only by the force between them. This force dominates, since it increases without limit, while the forces exerted by the other particles remain finite. So we can ignore all the other particles and consider the collision as if it were between two isolated mass points by the method already indicated in II § 3.8. We can, therefore, use the solution given there for the points 1 and 2, and, with its help, evaluate the other solution functions during the critical time interval.

A multiple collision may be similarly treated. Much work has been done on the rigorous mathematical investigation of these matters, but we shall not consider them any more closely, since the ratio of astronomical gain to mathematical expense seems to us to be small (as does, indeed, the ratio of mathematical gain to mathematical expense). Collisions are certainly sufficiently exceptional to be ignored in the investigation of the planetary system. If two planets actually should collide, however, they will not pass through each other without resistance. We need not, therefore, continue with these unrealistic considerations.

3. We will now make some inferences from the equations of motion,

$$m_i\ddot{\boldsymbol{r}}_i = \sum_{j=1}^{n} \boldsymbol{f}_{ji}, \qquad i = 1, 2, \ldots, n,$$

which will enable us to simplify them. First, by summation, we obtain

$$\sum_{i=1}^{n} m_i\ddot{\boldsymbol{r}}_i = \sum_{i=1}^{n}\sum_{j=1}^{n} \boldsymbol{f}_{ji} \equiv 0,$$

since, by the first force postulate, (II § 1.4) each pair of terms $\boldsymbol{f}_{ji}$ and $\boldsymbol{f}_{ij}$ cancel in the double summation.

We integrate the last equation with respect to time, and obtain the "law of the conservation of total momentum":

$$\sum_{i=1}^{n} m_i\dot{\boldsymbol{r}}_i(t) \equiv \mu\boldsymbol{a}$$

where

$$\mu = \sum_{i=1}^{n} m_i$$

is the total mass of the system, and $\boldsymbol{a}$ is some vector that is constant with respect to time. The result reads, therefore, the sum of all the momenta $m_i\boldsymbol{v}_i$ is constant with respect to time, or—the same thing in other words—the three components of the vector $\sum_{1}^{n} m_i\boldsymbol{v}_i$ are (first) integrals of the equations of motion. (See II § 3.1.)

Integrating once again, we obtain

$$\sum_{i=1}^{n} m_i\boldsymbol{r}_i(t) \equiv \mu(\boldsymbol{a}t + \boldsymbol{b}),$$

where $\boldsymbol{b}$ is another vector constant in time. The corresponding first integral follows by eliminating $\boldsymbol{a}$ from the last two vector equations to give

$$\sum_{i=1}^{n} m_i\{\boldsymbol{r}_i(t) - \boldsymbol{v}_i(t) \cdot t\} = \mu \boldsymbol{b}.$$

The components of the vector

$$\sum_{1}^{n} m_i(\boldsymbol{r}_i - \boldsymbol{v}_i t)$$

are first integrals of the equations of motion, if $\boldsymbol{r}_i$ and $\boldsymbol{v}_i$ are considered as independent variables.

4. We can formulate the last two laws in a somewhat different manner if we introduce the concept of the centre of mass (or "centre of gravity"). The centre of mass is defined as the point

$$\boldsymbol{r}_* = \frac{\sum_{1}^{n} m_i \boldsymbol{r}_i}{\sum_{1}^{n} m_i}.$$

Its position vector is, therefore, the arithmetic mean of all the position vectors $\boldsymbol{r}_i$, weighted with their respective masses m_i. Differentiating once and twice with respect to the time variable, and comparing with the momentum law we obtain:

$$\left.\begin{aligned} \ddot{\boldsymbol{r}}_* &= 0, \\ \dot{\boldsymbol{r}}_* &= \boldsymbol{a}, \\ \boldsymbol{r}_* &= \boldsymbol{a}t + \boldsymbol{b} \end{aligned}\right\}$$

—the absolute acceleration of the centre of mass is zero, its velocity is constant with respect to time, and it moves, in absolute space, along a straight line. That is the "law of the centre of gravity".

This law can be further generalized. Consider the following generalization of the equations of motion of 3(1):

$$m_i \ddot{\boldsymbol{r}}_i = \sum_{j=1}^{n} \boldsymbol{f}_{ji} + \boldsymbol{g}_i, \qquad i = 1, 2 \ldots n, \tag{1}$$

where the vector $\boldsymbol{g}_i$ is any force influencing the system from outside. Summing over i gives

$$\mu \ddot{\boldsymbol{r}}_* = \sum_{i=1}^{n} m_i \ddot{\boldsymbol{r}}_i = \sum_{i=1}^{n} \boldsymbol{g}_i \tag{2}$$

—the centre of mass moves as if the total mass of the system were concentrated at it, and as if all the external forces (or, indeed, all the forces) acted at that point.

If we write equation (2) in the form

$$\frac{\mathrm{d}}{\mathrm{d}t} \sum_{i=1}^{n} m_i \dot{\boldsymbol{r}}_i = \sum_{i=1}^{n} \boldsymbol{g}_i, \tag{3}$$

then it may also be interpreted in the following way: the derivative of the total momentum, with respect to the time, is equal to the sum of the external forces.

5. If, in the equations of motion

$$m_i \ddot{\boldsymbol{r}}_i = \sum_{j=1}^{n} \boldsymbol{f}_{ji}, \qquad i = 1, 2, \ldots, n,$$

we replace the position vector $\boldsymbol{r}_i$ by $\boldsymbol{r}_i + \boldsymbol{a}t + \boldsymbol{b}$, where $\boldsymbol{a}$ and $\boldsymbol{b}$ denote two arbitrary constant vectors, then we still obtain the original equations of motion. The additional terms on the left-hand side are eliminated by differentiation, while those on the right cancel, since, by the second and third force postulates of II § 1.1, the force $\boldsymbol{f}_{ji}$ depends on $\boldsymbol{r}_i$ and $\boldsymbol{r}_j$ only through the combination $(\boldsymbol{r}_j - \boldsymbol{r}_i)$. The equations of motion, therefore, are invariant with respect to translations with constant velocities. By § 4, therefore, we can so choose the coordinate system that the centre of mass shall always be at the origin:

$$\boldsymbol{r}_* \equiv 0.$$

This result is not a very surprising one. For a system of two gravitating particles the concepts were so formed that a result of this sort should be obtained, namely, that the system as a whole should be at rest in "absolute space". The new statement is more precise and more general: the centre of mass remains at rest.

6. A further conservation law, which is a generalization of the law of areas can be obtained by forming the quantity

$$\sum_{1}^{n} \boldsymbol{r}_i \times m_i \ddot{\boldsymbol{r}}_i$$

in the following manner:

$$\sum_{1}^{n} \boldsymbol{r}_i \times m_i \ddot{\boldsymbol{r}}_i = \sum_{i=1}^{n} \left(\boldsymbol{r}_i \times \sum_{j=1}^{n} \boldsymbol{f}_{ji}\right)$$
$$= \sum_{i=1}^{n} \sum_{j=1}^{n} \boldsymbol{r}_i \times \boldsymbol{f}_{ji}.$$

By exchanging the indices i and j in the factors $\boldsymbol{f}_{ji}$ we obtain

$$\sum_{i=1}^{n} \boldsymbol{r}_i \times m_i \ddot{\boldsymbol{r}}_i = -\sum_{i=1}^{n} \sum_{j=1}^{n} \boldsymbol{r}_i \times \boldsymbol{f}_{ij}$$

and, by exchanging the summation indices on the right-hand side of the same equation:

$$\sum_{i=1}^{n} \boldsymbol{r}_i \times m_i \ddot{\boldsymbol{r}}_i = \sum_{i=1}^{n} \sum_{j=1}^{n} \boldsymbol{r}_j \times \boldsymbol{f}_{ij}.$$

Addition of these last two equations gives

$$2\sum_{i=1}^{n} \boldsymbol{r}_i \times m_i \ddot{\boldsymbol{r}}_i = \sum_{i=1}^{n} \sum_{j=1}^{n} (\boldsymbol{r}_j - \boldsymbol{r}_i) \times \boldsymbol{f}_{ij}.$$

Now, by the second postulate of II § 1.1 the vectors $(\boldsymbol{r}_j - \boldsymbol{r}_i)$ and $\boldsymbol{f}_{ij}$ are collinear, so the right-hand side vanishes, and we obtain:

$$\sum_{i=1}^{n} \boldsymbol{r}_i \times m_i \ddot{\boldsymbol{r}}_i \equiv 0.$$

The left-hand side of this equation is the derivative, with respect to time, of the vector

$$\sum_{1}^{n} \boldsymbol{r}_i \times m_i \dot{\boldsymbol{r}}_i.$$

Therefore,

$$\frac{\mathrm{d}}{\mathrm{d}t} \sum_{i=1}^{n} \boldsymbol{r}_i \times m_i \dot{\boldsymbol{r}}_i \equiv 0,$$

and, integrating with respect to t, it follows that

$$\sum_{i=1}^{n} \boldsymbol{r}_i \times m_i \dot{\boldsymbol{r}}_i \equiv \textbf{const.}$$

This is the law of the conservation of the total moment of momentum. (The quantity $\boldsymbol{r}_i \times m_i \dot{\boldsymbol{r}}_i$ is called the moment of the momentum $m_i \dot{\boldsymbol{r}}_i$.) For $n = 2$ we obtain the law of areas if the law of the centre of gravity of section 4 is also taken into consideration.

7. The law of the conservation of moment of momentum may be generalized. Let the force $\boldsymbol{f}_i$ be composed of the internal forces $\boldsymbol{f}_{ij}$ and the external force $\boldsymbol{g}_i$, so that

$$\boldsymbol{f}_i = \sum_{j=1}^{n} \boldsymbol{f}_{ji} + \boldsymbol{g}_i \tag{1}$$

and

$$m_i \ddot{\boldsymbol{r}}_i = \sum_{j=1}^{n} \boldsymbol{f}_{ji} + \boldsymbol{g}_i. \tag{2}$$

Then, as in the previous section, we may obtain

$$\frac{\mathrm{d}}{\mathrm{d}t}\sum_{i=1}^{n}\boldsymbol{r}_i \times m_i\dot{\boldsymbol{r}}_i = \sum_{i=1}^{n} \boldsymbol{r}_i \times \boldsymbol{g}_i. \tag{3}$$

The right-hand side is the (total) moment of the external forces. Equation (1) can therefore be expressed as follows: the derivative of the total moment of momentum, with respect to time, is equal to the total moment of the external force. It is evident, from this form, that equation (3) is exactly analogous to the generalized law of momentum 4(3).

8. The last conservation law that we shall prove is the energy law. This states: if the equations of motion may be written in the form

$$m_i\ddot{x}_i = -\frac{\partial V}{\partial x_i}, \qquad i = 1, 2, \ldots, n, \text{ etc.},$$

where $V \equiv V(\boldsymbol{r}_1, \ldots, \boldsymbol{r}_n)$ is a function of the n position vectors $\boldsymbol{r}_1, \ldots, \boldsymbol{r}_n$ (called the "potential" of the "potential function" of the system), then the "energy function"

$$F \equiv \tfrac{1}{2}\sum_{i=1}^{n} m_i\boldsymbol{v}_i^2 + V(\boldsymbol{r}_1, \ldots, \boldsymbol{r}_n)$$

of the independent variables $\boldsymbol{r}_1, \ldots, \boldsymbol{r}_n, \boldsymbol{v}_1, \ldots, \boldsymbol{v}_n$ is a first integral of the equations of motion.

Proof: From

$$m_i\ddot{x}_i = -\frac{\partial V}{\partial x_i}, \quad \text{etc.}$$

it follows (similarly as in II § 3.3), by multiplying with $\dot{x}_i$ etc. and summing over i, that

$$\sum_{i=1}^{n} m_i(\ddot{x}_i\dot{x}_i + \ddot{y}_i\dot{y}_i + \ddot{z}_i\dot{z}_i) \equiv -\sum_{i=1}^{n}\left(\frac{\partial V}{\partial x_i}\dot{x}_i + \frac{\partial V}{\partial y_i}\dot{y}_i + \frac{\partial V}{\partial z_i}\dot{z}_i\right)$$

or

$$\frac{\mathrm{d}}{\mathrm{d}t}\left\{\tfrac{1}{2}\sum_{i=1}^{n} m_i(\dot{x}_i^2 + \dot{y}_i^2 + \dot{z}_i^2) + V\right\} \equiv 0$$

which is what was to be proved.

We shall now show that for a system of gravitating mass points the equations of motion can actually be written in the required form. Consider, for example, the first equation. Its right-hand side reads, by § 1.1

$$Gm_1 \sum_{j=2}^{n} m_j \frac{x_j - x_1}{|\boldsymbol{r}_j - \boldsymbol{r}_1|^3}.$$

We assert that this is the negative derivative of

$$V \equiv -\tfrac{1}{2}G \underbrace{\sum_{i=1}^{n} \sum_{j=1}^{n}}_{i \neq j} \frac{m_i m_j}{|\boldsymbol{r}_i - \boldsymbol{r}_j|}$$

with respect to x_1. In fact,

$$\begin{aligned}\frac{\partial V}{\partial x_1} &= -\tfrac{1}{2}G \frac{\partial}{\partial x_1}\left\{m_1 \sum_{j=2}^{n} \frac{m_j}{|r_1 - r_j|} + m_1 \sum_{i=2}^{n} \frac{m_i}{|\boldsymbol{r}_i - \boldsymbol{r}_1|}\right\} \\ &= -Gm_1 \sum_{j=2}^{n} m_j \frac{\partial}{\partial x_1} \frac{1}{|\boldsymbol{r}_1 - \boldsymbol{r}_j|} \\ &= Gm_1 \sum_{j=2}^{n} m_j \frac{1}{|\boldsymbol{r}_1 - \boldsymbol{r}_j|^2} \frac{\partial}{\partial x_1} |\boldsymbol{r}_1 - \boldsymbol{r}_j|.\end{aligned}$$

Since, now,

$$\frac{\partial}{\partial x_1} |\boldsymbol{r}_1 - \boldsymbol{r}_j| = \frac{x_1 - x_j}{|\boldsymbol{r}_1 - \boldsymbol{r}_j|},$$

it follows that

$$\frac{\partial V}{\partial x_1} = -Gm_1 \sum_{j=2}^{n} m_j \frac{x_j - x_1}{|\boldsymbol{r}_j - \boldsymbol{r}_1|^3},$$

as asserted. Corresponding proofs hold for the other equations.

9. Thus far we have obtained ten first integrals of the equations of motion. They are the energy integral F, the three integrals of moment of momentum $\sum \boldsymbol{r}_i \times m_i \boldsymbol{v}_i$, the three integrals of momentum $\sum m_i \boldsymbol{v}_i$, and three integrals of the centre of gravity $\sum m_i(\boldsymbol{r}_i - \boldsymbol{v}_i t_i)$. No further integrals are known, either in general, or in the particular case of Newtonian gravitation.

The use of these integrals lies in the fact that, for each one, the equation which expresses the corresponding conservation law can be solved for any one of the variables as a function of the others. This particular variable may then be eliminated from the equations of motion by substituting the function thus obtained. The number of differential equations and the number of unknown functions are then both reduced by one. If there were as many (independent) integrals as there are equations, the problem would be solved. Each integral, therefore, simplifies the problem a little.

On the other hand, the required solution of the integral, with respect to any variable, can lead to quite complicated working. In the mechanics of the planetary system, therefore, only six of the ten integrals are commonly used, namely, the integrals of momentum and of the centre of gravity. At this point it is convenient to return from absolute to relative coordinates, the coordinate origin once again being in the Sun. These coordinates are also more closely related to observable quantities. (We use the absolute space and the absolute coordinates, therefore, only in order to express the differential equations for relative motion.) This will be done in the following section.

10. The equation of motion of the Sun, in absolute coordinates, is

$$m_0 \ddot{\boldsymbol{r}}_0 = \sum_{j=1}^{n} \boldsymbol{f}_{j0}, \qquad m_0 = 1, \tag{1}$$

and that of any particular planet is

$$m_i \ddot{\boldsymbol{r}}_i = \boldsymbol{f}_{0i} + \sum_{j=1}^{n} \boldsymbol{f}_{ji}, \qquad i = 1, 2, \ldots, n. \tag{2}$$

The relative acceleration of the ith planet, with respect to the Sun, is thus given by

$$(\boldsymbol{r}_i - \boldsymbol{r}_0)^{\cdot\cdot} = \frac{\boldsymbol{f}_{0i}}{m_i} + \sum_{j=1}^{n} \left(\frac{\boldsymbol{f}_{ji}}{m_i} - \frac{\boldsymbol{f}_{j0}}{m_0} \right), \qquad i = 1, 2, \ldots, n. \tag{3}$$

In place of the original $(n + 1)$ vector equations of motion, we now have only n such equations. We should also note that the right-hand side depends only on the n vectors $(\boldsymbol{r}_i - \boldsymbol{r}_0)$. In fact, the forces $\boldsymbol{f}_{ji}$ depend only on the vectors $(\boldsymbol{r}_j - \boldsymbol{r}_i)$, and these can be represented in the form

$$\boldsymbol{r}_j - \boldsymbol{r}_i = (\boldsymbol{r}_j - \boldsymbol{r}_0) - (\boldsymbol{r}_i - \boldsymbol{r}_0),$$

so that the right-hand side can be considered as function of the vectors $(\boldsymbol{r}_i - \boldsymbol{r}_0)$ alone. The equations

$$\left.\begin{aligned} (\boldsymbol{r}_i - \boldsymbol{r}_0)^{\cdot} &= (\boldsymbol{v}_i - \boldsymbol{v}_0), \\ (\boldsymbol{v}_i - \boldsymbol{v}_0)^{\cdot} &= \frac{\boldsymbol{f}_{0i}}{m_i} + \sum_{j=1}^{n} \left(\frac{\boldsymbol{f}_{ji}}{m_i} - \frac{\boldsymbol{f}_{0j}}{m_0} \right), \end{aligned}\right\}$$
$$i = 1, 2, \ldots, n,$$

form a system of $6n$ ordinary differential equations, while the original system (1), (2) contained $6(n + 1)$ differential equations.

If the reduced system can be solved, then the absolute coordinates and velocities can be calculated by means of the laws of momentum and of the centre of gravity. There is in general, however, no need for this.

Our system of equations, when written out, reads

$$\left.\begin{aligned}
(\boldsymbol{r}_i - \boldsymbol{r}_0)^{\cdot} &= (\boldsymbol{v}_i - \boldsymbol{v}_0), \\
(\boldsymbol{v}_i - \boldsymbol{v}_0)^{\cdot} &= -Gm_0 \frac{\boldsymbol{r}_i - \boldsymbol{r}_0}{|\boldsymbol{r}_i - \boldsymbol{r}_0|^3} - \\
&\quad - G\sum_{j=1}^{n} m_j \frac{\boldsymbol{r}_j - \boldsymbol{r}_0}{|\boldsymbol{r}_j - \boldsymbol{r}_0|^3} - \\
&\quad - G\sum_{\substack{j=1 \\ i \neq j}}^{n} m_j \frac{\boldsymbol{r}_i - \boldsymbol{r}_j}{|\boldsymbol{r}_i - \boldsymbol{r}_j|^3}.
\end{aligned}\right\}$$

We introduce now the relative coordinates $\boldsymbol{r}_i' = \boldsymbol{r}_i - \boldsymbol{r}_0$. For simplicity, however, we omit the prime in $\boldsymbol{r}_i'$. Then we obtain:

$$\left.\begin{aligned}
\dot{\boldsymbol{r}}_i &= \boldsymbol{v}_i, \\
\dot{\boldsymbol{v}}_i &= -G(m_0 + m_i)\frac{\boldsymbol{r}_i}{|\boldsymbol{r}_i|^3} - \\
&\quad - G\sum_{\substack{j=1 \\ j \neq i}}^{n} m_j\left(\frac{\boldsymbol{r}_j}{|\boldsymbol{r}_j|^3} - \frac{\boldsymbol{r}_j - \boldsymbol{r}_i}{|\boldsymbol{r}_j - \boldsymbol{r}_i|^3}\right), \\
&\qquad i = 1, 2, \ldots, n.
\end{aligned}\right\} \tag{4}$$

These equations will be used in all that follows. Each single term of the summation expression represents the disturbing force exerted by the other planets. It is noteworthy that this force depends not only on the position of the disturbing planet, relative to the disturbed planet (expressed through $(\boldsymbol{r}_j - \boldsymbol{r}_i)$), but also on its position relative to the Sun (expressed through $\boldsymbol{r}_j$). The reason for this is the "absolute acceleration" of the coordinate origin—which coincides with the Sun. If we had tried to develop the dynamics of the planetary system directly from the relative Kepler motions, without introducing absolute space, these terms might easily have been overlooked. That they are correct is demonstrated by the agreement between computation and observation. We turn now to the integration of these equations.

2. Perturbations in the coordinates

1. From now on, for the sake of simplicity, we will consider only the case of two planets. The general case does not involve any new principle. If the position vectors of the two planets, as seen from the Sun, are denoted by $\boldsymbol{r}$ and $\boldsymbol{r}'$, the masses by m and m', and the masses combined with that of the Sun by μ and μ', then the equations of motion read:

$$\left.\begin{aligned} \ddot{\boldsymbol{r}} &= -G\mu\frac{\boldsymbol{r}}{|\boldsymbol{r}|^3} + Gm'\left(\frac{\boldsymbol{r}'-\boldsymbol{r}}{|\boldsymbol{r}'-\boldsymbol{r}|^3} - \frac{\boldsymbol{r}'}{|\boldsymbol{r}'|^3}\right), \\ \ddot{\boldsymbol{r}}' &= -G\mu'\frac{\boldsymbol{r}'}{|\boldsymbol{r}'|^3} + Gm\left(\frac{\boldsymbol{r}-\boldsymbol{r}'}{|\boldsymbol{r}-\boldsymbol{r}'|^3} - \frac{\boldsymbol{r}}{|\boldsymbol{r}|^3}\right). \end{aligned}\right\} \qquad (1)$$

They correspond to a system of 12 differential equations of the first order. To integrate them we apply the method of successive approximations of II § 2.2. We took there, as a zero order approximation, the constant initial conditions, but we noted that any other approximate solution may also be chosen and then improved, step by step. The iteration process, then, will in general converge still more quickly than with the constant zero approximation, since this is readily seen to be the worst of all the approximations used in practice.

For the three body problem discussed here we already know a good zero order approximation, namely, the Kepler motions that the two planets would describe if each was undisturbed by the other, that is to say if m or m' were, in turn, equal to zero. If we start from this approximation, we should quickly obtain much better approximate solutions. In fact, one iteration is often sufficient.

In the calculation of this first approximation, a special simplification appears, which is not possible for the higher order approximations. The first terms of the right-hand sides of equations (1) are simply the accelerations of the Kepler motions of the planets. If we take these terms to the left-hand sides, then we have the differential accelerations $(\Delta\boldsymbol{r})^{\cdot\cdot}$ and $(\Delta\boldsymbol{r}')^{\cdot\cdot}$ that must be added to the Kepler accelerations to obtain the first approximation. In the notation thus introduced, the equations read:

$$(\Delta \boldsymbol{r})^{\cdot\cdot} = Gm'\left(\frac{\boldsymbol{r}' - \boldsymbol{r}}{|\,\boldsymbol{r}' - \boldsymbol{r}\,|^3} - \frac{\boldsymbol{r}'}{|\,\boldsymbol{r}'\,|^3}\right), \quad \text{etc.,} \tag{2}$$

or

$$(\Delta \boldsymbol{r})^{\cdot\cdot} = \frac{m'}{\mu'}\,\ddot{\boldsymbol{r}}' + Gm'\,\frac{\boldsymbol{r}' - \boldsymbol{r}}{|\,\boldsymbol{r}' - \boldsymbol{r}\,|^3}, \quad \text{etc.,} \tag{3}$$

where the right-hand sides are now known functions of the time, which have to be twice integrated. The problem is essentially, therefore, to integrate

$$\frac{\boldsymbol{r}' - \boldsymbol{r}}{|\,\boldsymbol{r}' - \boldsymbol{r}\,|^3}.$$

The perturbations are reckoned as the deviations from the "osculating Kepler ellipse," that is, the ellipse that the disturbed planet would describe if it started from the initial conditions $\boldsymbol{v}_0$, $\boldsymbol{r}_0$, in the absence of the disturbing planet.

2. This integration can be performed numerically. It is also possible to expand the functions on the right-hand sides of the perturbation equations in Fourier series of the mean anomalies

$$M = n(t - T),$$

and

$$M' = n'(t' - T).$$

The series may then be integrated analytically with respect to the time, term by term. In order to write down the Fourier series, we use Euler's theorem:

$$\cos x \pm \sqrt{-1}\,\sin x = e^{\pm x\sqrt{-1}}.$$

The Fourier series of any (real) function $f(x)$ is

$$\tfrac{1}{2}a_0 + \sum_{\nu=1}^{\infty} (a_\nu \cos \nu x + b_\nu \sin \nu x), \tag{1}$$

where the coefficients are given by

$$\left.\begin{aligned} a_\nu &= \frac{1}{\pi}\int_0^{2\pi} f(y)\cos\nu y\,dy, \\ b_\nu &= \frac{1}{\pi}\int_0^{2\pi} f(y)\sin\nu y\,dy, \end{aligned}\right\} \qquad \nu = 0, 1, 2, \ldots . \tag{2}$$

We now set

$$\left.\begin{aligned} a_{-\nu} &= +a_\nu, \\ b_{-\nu} &= -b_\nu, \end{aligned}\right\} \qquad \nu = 0, 1, 2 \ldots \tag{3}$$

and

$$c_\nu = \tfrac{1}{2}(a_\nu + b_\nu\sqrt{-1}), \qquad \nu = 0, \pm 1, \pm 2, \ldots, \tag{4}$$

so that, as may easily be verified,

$$c_\nu e^{-\nu x\sqrt{-1}} + c_{-\nu}e^{+\nu x\sqrt{-1}} = a_\nu \cos\nu x + b_\nu \sin\nu x. \tag{5}$$

The Fourier series of $f(x)$ can, therefore, be written in the form

$$\sum_{\nu=-\infty}^{+\infty} c_\nu e^{-\nu x\sqrt{-1}}, \tag{6}$$

where, from (2), (3) and (4), it follows that

$$c_\nu = \frac{1}{2\pi}\int_0^{2\pi} f(x)e^{+\nu x\sqrt{-1}}\,dx, \qquad \nu = 0, \pm 1, \pm 2, \ldots \tag{7}$$

Similarly a function $f(x, y)$ of the two variables x and y has the Fourier series

$$\sum\sum_{\mu,\nu=-\infty}^{+\infty} c_{\mu\nu} e^{-(\mu x+\nu y)\sqrt{-1}}, \tag{8}$$

where

$$c_{\mu\nu} = \frac{1}{(2\pi)^2}\int_0^{2\pi}\int_0^{2\pi} f(x, y)e^{+(\mu x+\nu y)\sqrt{-1}}\,dx\,dy,$$

$$\begin{aligned} \mu &= 0, \pm 1, \pm 2, \ldots, \\ \nu &= 0, \pm 1, \pm 2, \ldots \end{aligned} \tag{9}$$

We apply this result to the function $(\boldsymbol{r}' - \boldsymbol{r})/|\,\boldsymbol{r}' - \boldsymbol{r}\,|^3$ of the two independent variables M and M'. We obtain the Fourier series

$$\sum_{\nu=-\infty}^{+\infty}\ \sum_{\nu'=-\infty}^{+\infty} \boldsymbol{c}_{\nu\nu'} \exp\{-(\nu M + \nu' M')\sqrt{-1}\}$$

where the coefficients $\boldsymbol{c}_{\nu\nu'}$, are given by

$$\boldsymbol{c}_{\nu\nu'} = \frac{1}{4\pi^2}\int\limits_0^{2\pi}\int\limits_0^{2\pi} \frac{\boldsymbol{r}' - \boldsymbol{r}}{|\,\boldsymbol{r}' - \boldsymbol{r}\,|^3} \exp\{+\sqrt{-1}(\nu M + \nu' M')\}\,\mathrm{d}M\,\mathrm{d}M'.$$

Now, as we know from I § 5.5, the variables M and M' do not appear explicitly in $(\boldsymbol{r}' - \boldsymbol{r})/|\,\boldsymbol{r}' - \boldsymbol{r}\,|^3$, but only through the eccentric anomalies E and E', which are related to M and M' by Kepler's equations

$$\left.\begin{aligned} E - e\sin E &= M, \\ E' - e'\sin E' &= M'. \end{aligned}\right\}$$

Then, if as in I § 5.3, we choose E and E' as integration variables, we obtain the following expression for the coefficients $\boldsymbol{c}_{\nu\nu'}$,:

$$\boldsymbol{c}_{\nu\nu'} = \frac{1}{4\pi^2}\int\limits_0^{2\pi}\int\limits_0^{2\pi} \frac{\boldsymbol{r}' - \boldsymbol{r}}{|\,\boldsymbol{r}' - \boldsymbol{r}\,|^3} \exp\{+\sqrt{-1}(\nu E + \nu' E' - \nu e\sin E - \\ - \nu' e'\sin E')\}\,.\,(1 - e\cos E)(1 - e'\cos E')\,\mathrm{d}E\,\mathrm{d}E'.$$

It is probably best to calculate these coefficients numerically. Usually, however, they have been represented as power series in the eccentricities, or in the ratio of the semi-major axes, or in some other parameter. This seems inconvenient to us: the chief difficulty is the denominator $|\,\boldsymbol{r}' - \boldsymbol{r}\,|^3$. It renders necessary a large number of elementary calculations in any series development of the Fourier coefficients. These calculations teach us nothing about either astronomy or mathematics, and only obscure the whole problem.

If we now set

$$\left.\begin{aligned} M &= n(t - T), \\ M' &= n'(t - T'), \end{aligned}\right\}$$

then $(\boldsymbol{r}' - \boldsymbol{r})/|\,\boldsymbol{r}' - \boldsymbol{r}\,|^3$ can be expressed as a function of time through the series

$$\sum_{\nu,\nu'=-\infty}^{+\infty}\!\!\sum \boldsymbol{c}_{\nu\nu'},\exp\{\sqrt{-1}(\nu nT + \nu' n' T)'\}\exp\{-\sqrt{-1}(\nu n + \nu' n')t\}. \tag{10}$$

Now $\ddot{\boldsymbol{\rho}}(t) \equiv (\boldsymbol{r}' - \boldsymbol{r})/|\,\boldsymbol{r}' - \boldsymbol{r}\,|^3$ is an analytic function of t, and so the series converges absolutely and uniformly to $\ddot{\boldsymbol{\rho}}$, for all values of t. Hence the series that are obtained from (10) by first and second integrations with respect to the time variable t, also converge to

$$\dot{\boldsymbol{\rho}}(t) = \int_0^t \ddot{\boldsymbol{\rho}}(t)\, dt,$$

and

$$\boldsymbol{\rho}(t) = \int_0^t \dot{\boldsymbol{\rho}}(t)\, dt.$$

3. From the mathematical point of view, two cases must be distinguished in these integral series, according as the mean motions of the two planets are commensurable or not. If they are incommensurable, $\ddot{\boldsymbol{\rho}}$ is an almost periodic function of t, the integral of which is given by

$$\dot{\boldsymbol{\rho}}(t) = \boldsymbol{c}_{00}t + \sum_{\nu^2+\nu'^2>0}\sum \boldsymbol{c}_{\nu\nu'} \exp\{(\nu nT + \nu'n'T')\sqrt{-1}\} \times \frac{\exp\{-(\nu n + \nu'n')t\sqrt{-1}\} - 1}{-(\nu n + \nu'n')\sqrt{-1}}. \qquad (1)$$

In this expression arbitrarily small denominators $\nu n + \nu'n'$ appear, since the irrational number n'/n may be approximated, with arbitrary accuracy, by quotients of the form $-\nu/\nu'$. There can and must be, therefore, infinitely many terms with high numbers ν, ν' which have small coefficients in the series for $\ddot{\boldsymbol{\rho}}$, while in the corresponding series for $\dot{\boldsymbol{\rho}}$ they suddenly appear with very large coefficients. In the series for $\boldsymbol{\rho}$ itself, the small denominators are squared. Nevertheless, on account of its uniform convergence, the series can still represent the function, with any predetermined accuracy, if it is truncated to some appropriate length. The maximum error involved in this process is also known. If n'/n is rational, then each term of the series for $\dot{\boldsymbol{\rho}}$ for which $\nu n + \nu'n' = 0$ must be replaced by

$$c_{\nu\nu'} \exp\{-(\nu nT + \nu'n'T')\sqrt{-1}\}\,.\,t,$$

which can be regarded as the limit of the general term in the series (1) for

$$(\nu n + \nu'n') \to 0.$$

The representation of $\boldsymbol{\rho}$ is dealt with in a similar way.

In astronomical applications, the distinction between the two cases is, of course, unimportant. The fundamental fact, for such applications, is the uniformity of the convergence, which holds in both cases. It is more important whether the small demoninators appear with small values of ν, ν' or with large, since, in the first case, the corresponding terms in the truncated and integrated series become of considerable weight. This only becomes true after some time, however, that is when

$$| \exp (-[\nu n + \nu' n']t\sqrt{-1}) - 1 | \gg | \nu n + \nu' n' |.$$

Such a case is provided, for example, by the periods of Jupiter and Saturn, 11.862 and 29.457. The ratio of these two numbers is nearly as 2 : 5. For $\nu = 2$, $\nu' = -5$, therefore,

$$\begin{aligned} \nu n + \nu' n' &= \frac{2 \,.\, 2\pi}{11{\cdot}862} - \frac{5 \,.\, 2\pi}{29{\cdot}457} = -\frac{2\pi}{882{\cdot}5} \\ &= -0{\cdot}007418, \end{aligned}$$

so that, after the second integration, the coefficient $c_{2,-5}$ contains a factor of the order of 2×10^4. On the other hand, for the term $c_{2,5}$ the corresponding factor is

$$\left(2 \,.\, \frac{2\pi}{11{\cdot}86} + 5 \,.\, \frac{2\pi}{29{\cdot}46}\right)^{-2} = 0{\cdot}221,$$

and the integration does not affect the order of magnitude. The term (2, − 5) is called a long period term; terms like (2, 5) are short period terms. The period associated with (2, − 5) is 882·5 years. Hence, the influence of this term is appreciable only after some hundreds of years.

The term $c_{00}t$ in the series for $\dot{\rho}(t)$—and, in general, any term that is not bounded if $t \to \pm\infty$—is called a secular term. It is often supposed that if the secular term is absent and only trigonometric terms appear, then the function $\dot{\rho}(t)$ is bounded. This is certain to be true, however, only if the mean motions are commensurable, so that the trigonometric series represents a periodic function. If the mean motions are not commensurable, then the trigonometric series represents an almost periodic function, the integral of which may also be an almost periodic function, but is not necessarily so. If, and only if, it is an almost periodic function, is it bounded.

The distinction between these two cases, also, is pointless in astronomical applications. We deal with time intervals of only limited length. It is not important whether the coefficient c_{00}

is exactly zero, for example, but only whether it is so large that the function $\dot{\boldsymbol{\rho}}(t)$ must be considered, in the given interval, as a linear function on which small oscillations are superimposed, or whether these oscillations are so large that the linear terms are submerged by them. Moreover, all this work, at present, leads only to a first approximation. The analytical character of the rigorous solution cannot, in principle, be known from its first approximation, and is, in fact, still unknown to-day.

4. The secular term

$$m\boldsymbol{c}_{00} = \frac{Gmm'}{4\pi^2}\int_0^{2\pi}\int_0^{2\pi} \frac{\boldsymbol{r}' - \boldsymbol{r}}{|\,\boldsymbol{r}' - \boldsymbol{r}\,|^3}\,\mathrm{d}M\,\mathrm{d}M' \tag{1}$$

in the Fourier development of the disturbing force $Gmm'(\boldsymbol{r}' - \boldsymbol{r})/|\,\boldsymbol{r}' - \boldsymbol{r}\,|^3$ was interpreted by Gauss in the following intuitive manner. We denote by $\mathrm{d}s$ (or $\mathrm{d}s'$, respectively) the element of arc length on the undisturbed Kepler ellipse. The circumference of the ellipse is denoted by l, and the mean orbital velocity of the planet by $\bar{v} = l/P$. We now suppose each arc element $\mathrm{d}s$ to have a mass $m\dfrac{\mathrm{d}M}{2\pi}$, so that its "length density" is, therefore

$$\sigma = \frac{m}{2\pi}\frac{\mathrm{d}M}{\mathrm{d}s}. \tag{2}$$

This expression can also be written:

$$\sigma = \frac{m}{2\pi}\cdot\frac{2\pi}{P}\frac{\mathrm{d}t}{\mathrm{d}s} = \frac{m}{P}\frac{\mathrm{d}t}{\mathrm{d}s}$$

$$= \frac{m}{l}\frac{\mathrm{d}t}{\mathrm{d}s/\bar{v}}. \tag{3}$$

The local length density, σ, is therefore equal to the mean length density, m/l, multiplied by the ratio of the actual "time of passage," $\mathrm{d}t$, of the planet, through the element of arc, $\mathrm{d}s$, to the mean time of passage, $\mathrm{d}s/\bar{v}$.

By (2), the secular term (1) can be written:

$$G\int_0^{l}\int_0^{l'}\frac{\boldsymbol{r}' - \boldsymbol{r}}{|\,\boldsymbol{r}' - \boldsymbol{r}\,|^3}\,\sigma(\boldsymbol{r})\,\mathrm{d}s\,.\,\sigma(\boldsymbol{r}')\,\mathrm{d}s'.$$

This expression can be thought of in the following way: the secular component of the disturbing force between the two planets is the sum of all the forces that each pair of elements in two material rings, with length densities $\sigma(\boldsymbol{r})$ and $\sigma'(\boldsymbol{r}')$, would exert on each other.

5. The methods of §§ 1 and 2 can be used to solve the problem of finding the orbital elements of a still unknown planet (Neptune) from the perturbations of a known one (Uranus). From the results of I § 2.4, we would take the eccentricity and the inclination of the unknown orbit to be vanishingly small. The remaining unknowns are then the mass m', the semi-major axis a', and—in place of the time of perihelion—the direction θ_0', of the supposed planet, at the time $t = 0$. The elements of the osculating orbital ellipse of Uranus are known at the time $t = 0$, as also are the direction $\boldsymbol{u}$ and the distance ϱ, of the planet, as seen from the Earth. These last two quantities refer to Uranus as it would be if it were not disturbed, (or, alternatively if it were disturbed only by Jupiter and Saturn). The deviation of its actual geocentric direction from the undisturbed value is denoted by $\Delta\boldsymbol{u}$.

Now let λ and β be the geocentric longitude and latitude, which are two coordinate directions for the observer on the Earth, so that

$$\boldsymbol{u} = \begin{pmatrix} \cos\beta\cos\lambda \\ \cos\beta\sin\lambda \\ \sin\beta \end{pmatrix}, \tag{1}$$

and $\beta = 0$ for all points in the orbital plane of the Earth. We also write

$$\frac{\partial\boldsymbol{u}}{\partial\lambda} \equiv \boldsymbol{u}_\lambda \cos\beta, \quad \frac{\partial\boldsymbol{u}}{\partial\beta} \equiv \boldsymbol{u}_\beta. \tag{2}$$

We have, as in II § 4,

$$\boldsymbol{r} = \boldsymbol{q} + \varrho\boldsymbol{u}, \tag{3}$$

for the position vector $\boldsymbol{r}(t)$ of Uranus, undisturbed by the unknown planet Neptune, where $\boldsymbol{q}$ denotes the position vector of the Earth.

For the position of Uranus disturbed by Neptune, we have, in the first approximation,

$$\begin{aligned}\Delta\boldsymbol{r} &= \Delta(\varrho\boldsymbol{u}) \\ &= \Delta\varrho\boldsymbol{u} + \varrho(\boldsymbol{u}_\lambda \cos\beta\Delta\lambda + \boldsymbol{u}_\beta\Delta_\beta),\end{aligned}$$

where $\Delta \boldsymbol{r}$ is the correction of $\boldsymbol{r}(t)$ for the perturbation by Neptune, etc. By scalar multiplication with $\boldsymbol{u}_\lambda$ we obtain:

$$\boldsymbol{u}_\lambda \Delta \boldsymbol{r} = \varrho \cos \beta \Delta \lambda. \tag{4}$$

Now, by 1(3), $\Delta \boldsymbol{r}$ is given by

$$\Delta \boldsymbol{r} = m' \left[\frac{\boldsymbol{r}'}{a'^3} + 4\pi^2 \int\!\!\int \frac{\boldsymbol{r}' - \boldsymbol{r}}{|\,\boldsymbol{r}' - \boldsymbol{r}\,|^3} \,\mathrm{d}t \,\mathrm{d}t \right], \tag{5}$$

where $\boldsymbol{r}'$ is the (undisturbed) position vector of Neptune. The vector $\boldsymbol{r}'$ is a function of a', θ_0' and t. We write

$$\left.\begin{aligned} a' &= \mathring{a}' + \delta a', \\ \theta_0' &= \mathring{\theta}_0' + \delta \theta_0', \\ m' &= \mathring{m}' + \delta m', \end{aligned}\right\} \tag{6}$$

where $\mathring{a}'$, $\mathring{\theta}_0'$ and $\mathring{m}'$ are any estimated values of a', θ_0' and m', and $\delta a'$, $\delta\theta_0'$ and $\delta m'$ are the required improvements. If we expand the right-hand side of (5) in powers of $\delta a'$, $\delta\theta_0'$ and $\delta m'$, and then terminate the series at the linear terms, we obtain for $\Delta \boldsymbol{r}$ an expression of the form

$$\Delta \boldsymbol{r} = \boldsymbol{b}_0 + \boldsymbol{b}_1 \,\delta a' + \boldsymbol{b}_2 \,\delta \theta_0' + \boldsymbol{b}_3 \,\delta m'. \tag{7}$$

The vectors $\boldsymbol{b}_0$, $\boldsymbol{b}_1$, $\boldsymbol{b}_2$, $\boldsymbol{b}_3$ depend, in a known way, on the elements of Uranus, on $\mathring{a}'$, $\mathring{\theta}_0'$, $\mathring{m}'$, as well as on t. By the method of section 2 we can develop these four vector functions in Fourier series with respect to t, the coefficients of which still depend only on numerically known quantities, and can therefore be numerically calculated. The $\boldsymbol{b}$'s themselves are therefore explicitly known functions of t. We introduce the expression (7), for $\Delta \boldsymbol{r}$, into the equation (4). All the quantities in (4), except $\delta a'$, $\delta\theta_0'$, $\delta m'$ are now known: $\boldsymbol{u}_\lambda$ and ϱ from the osculating undisturbed motion of Uranus, and $\Delta\lambda(t)$ from the observations. The three unknown quantities can be evaluated by the method of least squares. This process can be iterated.

How are the first approximations obtained? For the semi-axis of Neptune, a', both Adams and Leverrier chose a value corresponding to the Titius–Bode rule (I § 2.4.) Unfortunately, however, this rule fails just for Neptune, and, in fact, both investigators gave final values for a' which were considerably too large. There are other indications of the required approximations. If we neglect the eccentricity and inclination of the orbit of Uranus, then the perturbation of its mean anomaly, Δw, should be a periodic function of the time that nearly coincides with $\Delta\lambda$. The period is the time between two successive conjunctions of the two planets (the synodic period of Neptune with respect to Uranus). It is determined, essentially, by the semi-axis a'. Thus a' may be estimated from it. Further, the form of the curve "$\Delta\lambda$ versus t" depends, primarily, only on the semi-axis a'. That is immediately and intuitively obvious: the nearer a' is to a, the stronger the disturbing influence, but, also, the more strictly localized in time is its action. A variation of the unknown mass m', on the other hand, only produces an alteration of scale in $\Delta\lambda$, since $\Delta\lambda$ and m' are proportional to each other. Lastly, θ' serves only to specify the phase of the periodic function $\Delta\lambda$. An analysis of the curve of "$\Delta\lambda$ versus t" would, if it were known over a whole synodic period, provide some indication of the first approximate values of a', m', and θ'_0. Only one third of such a period was available to Adams and Leverrier, however, and they would have had to try several sets of plausible values of a', m', θ', had they used the method just sketched. We gave this method here to show that, in principle, the problem is elementary.

In the successive improvements of the initial values, the simplifying assumptions about the position and shape of the orbit may be dropped, and an attempt made to determine all six elements from the observations. It is readily seen, however, that the elements neglected above can (except in very extreme cases) be given only very inaccurately (cf. I § 6.6), so that the much greater expense of calculation is scarcely rewarded. Adams and Leverrier actually did determine all six elements, but (as was to be expected) for those particular elements only rather inaccurate values were obtained.

6. In § 1 we have calculated the "perturbations in the coordinates" of the position vectors. Instead of these we can also consider the "perturbations in the elements": we no longer suppose the six elements of a planetary orbit to be constant in time, but to be quantities which themselves change slowly, under the influence of the perturbations. The planets move along their respective "osculating ellipses", which, however, are

no longer fixed, but at each instant assume a new position, size and shape.

For the calculation of perturbations in the elements, we refer back to I § 6.2. The calculation becomes somewhat clearer if we use the dynamical results of II § 3. In II § 3.7, for example, we found the expression

$$a = -\frac{G\mu}{2h} \tag{1}$$

for the semi-major axis a, where h, the energy constant, is given by

$$h = \tfrac{1}{2}\boldsymbol{v}^2 - \frac{G\mu}{|\boldsymbol{r}|}, \tag{2}$$

as in II § 3.3. We denote by Δ the perturbations of the Kepler motion. In the first approximation, then, by (1),

$$\frac{\Delta a}{a} = -\frac{\Delta h}{h} \tag{3}$$

and by (2),

$$\Delta h = \boldsymbol{v}\Delta\boldsymbol{v} + \frac{G\mu}{|\boldsymbol{r}|^3}\boldsymbol{r}\Delta\boldsymbol{r}, \tag{4}$$

where $\Delta\boldsymbol{v}$ and $\Delta\boldsymbol{r}$ are as given in § 1.

Similarly, we obtain from

$$\boldsymbol{c} = \tfrac{1}{2}(\boldsymbol{r} \times \boldsymbol{v}) \tag{5}$$

(II § 3.1) the relation

$$\Delta\boldsymbol{c} = \tfrac{1}{2}(\Delta\boldsymbol{r} \times \boldsymbol{v}) + \tfrac{1}{2}(\boldsymbol{r} \times \Delta\boldsymbol{v}). \tag{6}$$

By means of the formulae for Ω, ι and e in II § 3.2, the perturbations $\Delta\Omega$, $\Delta\iota$ and Δe can now be easily calculated. (The perturbation Δa is here assumed to be already known.)

The secular perturbations in the longitude and time of perihelion can, finally, be obtained in a similar, but more complicated manner from the formulae of I § 6.3 and 2.

In all this work we have used only the first approximations for the perturbations. Naturally there is no difficulty of principle in computing approximations of higher orders.

Instead of evaluating the perturbations of the elements from those of the coordinates, they can also be evaluated directly by setting up and integrating differential equations for the disturbed elements. This method is the simpler and more usual. We shall describe it in the following two sections; first quite generally for an arbitrary system of differential equations, and then for those of celestial mechanics.

3. Perturbations in the elements

1. Let the system of differential equations

$$\dot{x}_i = f_i(\mathbf{x}, t), \qquad i = 1, 2, \ldots, n, \tag{1}$$

for the n-dimensional vector $\mathbf{x} = (x_1, \ldots, x_n)$, have the general solution

$$x_i = \phi_i(\mathring{\mathbf{x}}, t), \qquad i = 1, 2, \ldots, n. \tag{2}$$

Here $\mathring{\mathbf{x}}$ is the given, arbitrary value of $\mathbf{x}$ at the time $t = 0$, so that

$$\mathring{x}_i = \phi_i(\mathring{\mathbf{x}}, 0).$$

Our problem now is to solve the "neighbouring" system of equations,

$$\dot{x}_i = f_i(\mathbf{x}, t) + g_i(\mathbf{x}, t), \qquad i = 1, 2, \ldots, n, \tag{3}$$

if the general solution (2) of the system (1) is assumed to be known. It is assumed that the additional terms g_i are (absolutely) small. It can at once be proved by means of successive approximations that the solutions of both systems of equations differ from each other, in a narrow time interval, by an arbitrarily small amount, if $|\, g \,|$ is sufficiently small. In other words, the solution depends "continuously" on the right-hand side. It also depends continuously on the initial vector $\mathring{\mathbf{x}}$.

We now attempt to solve the neighbouring system by substituting an unknown function $\psi_i(\mathring{\mathbf{x}}, t)$ for the constants of integration, $\mathring{\mathbf{x}}$, in the solution, ϕ_i, of the "undisturbed system". Thus,

$$x_i = \phi_i\{\boldsymbol{\psi}(\mathring{\mathbf{x}}, t), t\}.$$

The functions ψ_i have to fulfil two conditions. They must reduce to $\mathring{\mathbf{x}}$ when $t = 0$, and they must satisfy the system of differential equations that is obtained by setting the trial solution $x_i = \phi_i(\boldsymbol{\psi}, t)$ into the above system of equations

$$\dot{x}_i = f_i + g_i.$$

The new system is

$$\sum_{j=1}^{n} \frac{\partial}{\partial \psi_j} \phi_i(\boldsymbol{\psi}, t) \cdot \frac{\partial \psi_j(\mathring{\boldsymbol{x}}, t)}{\partial t} + \frac{\partial \phi_i(\boldsymbol{\psi}, t)}{\partial t}$$
$$= f_i\{\boldsymbol{\phi}(\boldsymbol{\psi}, t), t\} + g_i\{\boldsymbol{\phi}(\boldsymbol{\psi}, t), t\}, \qquad i = 1, \ldots, n.$$

Now, by the definition of the function ϕ_i,

$$\frac{\partial \phi_i}{\partial t} = f_i(\boldsymbol{\phi}, t)$$

(for all $\mathring{\boldsymbol{x}}$, t; therefore, also for all $\boldsymbol{\psi}(\mathring{\boldsymbol{x}}, t), t)$, so the function $\psi_i(\mathring{\boldsymbol{x}}, t)$ is given by the solution of the system of differential equations

$$\sum_{j=1}^{n} \frac{\partial \phi_i}{\partial \psi_j} \dot{\psi}_j = g_i\{\boldsymbol{\phi}(\boldsymbol{\psi}, t), t\},$$

with the initial conditions

$$\psi_i(\mathring{\boldsymbol{x}}, t) = \mathring{x}_i.$$

To bring this system to the normal form we must solve it for the $\dot{\psi}_i$. This is possible, since the "functional determinant"

$$\left| \frac{\partial \phi_i(\mathring{\boldsymbol{x}}, t)}{\partial \mathring{x}_j} \right|$$

is different from zero. If $t = 0$, it is equal to one because of the relation

$$\phi_i(\mathring{\boldsymbol{x}}, t) = \mathring{x}_i.$$

The determinant is positive, therefore, within a certain neighbourhood around the time zero point. It can even be proved to be positive for all values of t, if there are no collisions.

The equations for the $\dot{\psi}_i$ are linear and homogeneous in the g_i. If these functions are small, therefore, so also are the $\dot{\psi}_i$. The method of successive approximations of II § 2.1 will, therefore, yield good approximations after only a few steps.

This method of the "variation of parameters" can be applied in exactly the same manner if the general solution has the form

$$x_i = \chi_i(\boldsymbol{C}, t),$$

where

$$\boldsymbol{C} = (C_1, \ldots, C_n)$$

are any constants of integration.

2. The method just described is the variation of parameters as given by Lagrange. Poisson gave a dual method. It is assumed that the solutions of the undisturbed system

$$x_i = f_i(\boldsymbol{x}, t), \qquad i = 1, 2, \ldots, n, \tag{1}$$

are given implicitly through first integrals $X_i(\boldsymbol{x}, t)$. Thus, if $\chi_i(\boldsymbol{C}, t)$ is the general solution of (1),

$$X_i[\boldsymbol{\chi}(C, t), t] \equiv C_i, \qquad i = 1, 2, \ldots, n. \tag{2}$$

For the disturbed system

$$\dot{x}_i = f_i(\boldsymbol{x}, t) + g_i(\boldsymbol{x}, t), \qquad i = 1, 2, \ldots, n, \tag{3}$$

we assume the solution to have the same form, but the C_i now depend on the time. To make a clear distinction between the functions $C_i = C_i(t)$ and the constants C_i, we shall henceforth denote the latter by $\mathring{C}_i$.

We now define the functions $X_{i0}(\boldsymbol{x}, t)$ and $X_{ij}(\boldsymbol{x}, t)$ by

$$\left.\begin{aligned} X_{i0}(\boldsymbol{x}, t) &\equiv \frac{\partial}{\partial t} X_i(\boldsymbol{x}, t), \\ X_{ij}(\boldsymbol{x}, t) &\equiv \frac{\partial}{\partial x_j} X_i(\boldsymbol{x}, t). \end{aligned}\right\} \tag{4}$$

Then, by differentiating the identities (2) with respect to t, we obtain,

$$\dot{C}_i(t) = X_{i0}[\boldsymbol{\chi}(C, t), t] + \sum_{j=1}^{n} X_{ij}[\boldsymbol{\chi}(C, t), t] \frac{\mathrm{d}}{\mathrm{d}t} \chi_j(C, t), \tag{5}$$

or, using equation (3),

$$\dot{\boldsymbol{C}}_i(t) = X_{i0}[\boldsymbol{\chi}(C, t), t] + \sum_{j=1}^{n} X_{ij}[\boldsymbol{\chi}(C, t), t][f_j(\boldsymbol{\chi}, t) + g_j(\boldsymbol{\chi}, t)]. \tag{6}$$

In particular, we have, for the undisturbed system (1),

$$0 = X_{i0}[\boldsymbol{\chi}(\mathring{C}, t), t] + \sum_{j=1}^{n} X_{ij}[\boldsymbol{\chi}(\mathring{C}, t), t] \,.\, f_j[\boldsymbol{\chi}(\mathring{C}, t), t] \tag{7}$$

for all $\mathring{\boldsymbol{C}}$. Therefore (7) still holds if $\boldsymbol{C}(t)$ is substituted for $\mathring{\boldsymbol{C}}$:

$$0 = X_{i0}[\boldsymbol{\chi}(C, t), t] + \sum_{j=1}^{n} X_{ij}[\boldsymbol{\chi}(C, t), t] \,.\, f_j[\boldsymbol{\chi}(C, t), t]. \tag{8}$$

Subtracting (8) from (6) we obtain

$$\boxed{\dot{C}_i = \sum_{j=1}^{n} \left(\frac{\partial X_i}{\partial x_j} g_j\right)_{x=\chi(C,\, t).}} \tag{9}$$

This system of equations for the disturbed elements $C_i(t)$ is fundamental to our treatment of perturbation problems. The equation for the disturbed elements can be written down at once if the solutions $\chi_i(C, t)$ and their inverse functions, the first integrals $X_i(x, t)$, are known for the undisturbed problem—as they are for the undisturbed two body problem.

3. The solution of a neighbouring system of equations can also be found by means of the "method of perturbations" in a narrower sense. Let the system of equations be

$$\dot{x}_i = f_i(\mathbf{x}, t, \varepsilon), \qquad i = 1, 2, \ldots, n,$$

where ε is a parameter. We assume that the solution the system of equations

$$\dot{x}_i = f_i(\mathbf{x}, t, 0), \qquad i = 1, 2, \ldots, n$$

is known, and that the functions f_i may be developed in power series of the variables x_i and of ε. We write the required solution as a power series in ε, namely:

$$x_i = \phi_i^{(0)}(\mathring{\mathbf{x}}, t) + \varepsilon\phi_i^{(1)}(\mathring{\mathbf{x}}, t) + \varepsilon^2\phi_i^{(2)}(\mathring{\mathbf{x}}, t) + \ldots,$$

and introduce this into the system of differential equations. Comparing the coefficients of like powers of ε on both sides, we obtain

$$\left.\begin{aligned} \dot{\phi}_i^{(0)} &= f_i(\boldsymbol{\phi}^{(0)}, t, 0), \\ \dot{\phi}_i^{(1)} &= \sum_{j=1}^{n} \frac{\partial f_i(\boldsymbol{\phi}^{(0)}, t, 0)}{\partial \phi_j^{(0)}}\, \phi_j^{(1)} + \left(\frac{\partial f_i(\boldsymbol{\phi}^{(0)}, t, \varepsilon)}{\partial \varepsilon}\right)_{\varepsilon=0}, \\ &\ldots\ldots\ldots \end{aligned}\right\}$$

while the initial conditions are,

$$\left.\begin{aligned} \phi_i^{(0)}(\mathring{\mathbf{x}}, 0) &= \mathring{x}_i, \\ \phi_i^{(1)}(\mathring{\mathbf{x}}, 0) &= 0, \\ \phi_i^{(2)}(\mathring{\mathbf{x}}, 0) &= 0, \\ &\ldots\ldots \end{aligned}\right\}$$

The solutions of the system of equations for $\phi_i^{(0)}$ are known, by our assumptions. The system of equations for $\phi_i^{(1)}$ is linear and inhomogeneous; its coefficients are already known functions. The same holds also for the systems of equations for $\phi_i^{(2)}$, $\phi_i^{(3)}$ etc., which are all of the form

$$\dot{\phi}_i^{(\lambda)} = \sum_{j=1}^{n} \frac{\partial f_i(\boldsymbol{\phi}^{(0)}, t, 0)}{\partial \phi_j^{(0)}} \phi_j^{(\lambda)} + r_i^{(\lambda)}(t).$$

The $r_i^{(\lambda)}(t)$ are known functions of the time variable, t, which are uniquely determined by the solutions $\phi_i^{(1)}, \phi_i^{(2)}, \ldots, \phi_i^{(\lambda-1)}$ of the preceding equations of the system. This can be proved by the method of mathematical induction.

This fact is of considerable practical importance in the computation of the functions ϕ_i. For suppose that the general solution of the corresponding homogeneous system of equations,

$$\dot{\phi}_i^{*} = \sum_{j=1}^{n} \frac{\partial f_i(\boldsymbol{\phi}^{(0)}, t, 0)}{\partial \phi_j^{(0)}} \phi_j^{*},$$

is known. Then, as will be shown in 7, the general solution of the inhomogeneous system may be obtained by simple quadratures. The chief difficulty is finding the solution of the homogeneous system. The above theorem, however, shows that this difficulty appears only once, the homogeneous system being the same for all n-tuples $\phi_i^{(1)}, \phi_i^{(2)}, \ldots$.

The system of equations for $\phi_i^{(1)}$, $\phi_i^{(2)}$, etc., becomes particularly simple if

$$f_i(x, t, 0) \equiv 0.$$

For, then

$$\left.\begin{aligned} \phi_i^{(0)} &= \mathring{x}_i, \\ \phi_i^{(1)} &= \int_0^t \left(\frac{\partial f(\mathring{x}, \tau, \varepsilon)}{\partial \varepsilon}\right)_{\varepsilon=0} d\tau, \\ &\cdots\cdots \end{aligned}\right\}$$

—all the functions ϕ_i may be evaluated by quadratures.

The convergence of all these series, for sufficiently small $|\varepsilon|$ can be proved, for example, by successive approximations.

This method can be applied to the original system of equations, or to that given by the variation of constants, if it contains a suitable small parameter. In celestial mechanics, the masses of the planets, for example, are such parameters. In practical applications the series are often broken off after the linear term. The method can be generalized when there are several (small) parameters.

4. The method of computing perturbations just explained can be used, even if the right-hand sides of the system of differential equations do not depend on any parameter. Consider the system

$$\dot{x}_i = f_i(\boldsymbol{x}, t), \qquad i = 1, 2, \ldots, n,$$

of which the particular solution

$$x_i = \phi_i^{(0)}(t)$$

is known, whose initial values are given by, for example,

$$\phi_i^{(0)}(0) = 0.$$

For the neighbouring initial values

$$\mathring{x}_i = \varepsilon a_i, \qquad a_i = \text{const.},$$

we again try a power series in ε as the solution. The working is quite analogous to that of the last section. We have to set $\partial f_i / \partial \varepsilon \equiv 0$. Thus we obtain the homogeneous linear system of equations,

$$\dot{\phi}_i^{(1)} = \sum_{j=1}^{n} \frac{\partial f_i(\boldsymbol{\phi}^{(0)}, t)}{\partial \phi_j^{(0)}} \phi_j^{(1)},$$

for the functions $\phi_i^{(1)}$, with the initial conditions

$$\phi_i^{(1)}(0) = a_i.$$

These equations are called Jacobi variational equations. For the following iterations we again obtain an inhomogeneous linear system of the form

$$\dot{\phi}_i^{(\lambda)} = \sum_{j=1}^{n} \frac{\partial f_i(\boldsymbol{\phi}^{(0)}, t)}{\partial \phi_j^{(0)}} \phi_j^{(\lambda)} + r_i^{(\lambda)}(t), \qquad \lambda = 2, 3, \ldots,$$

with the inital conditions

$$\phi_i^{(\lambda)}(0) = 0, \qquad \lambda = 2, 3 \ldots .$$

5. We now turn to linear systems of equations, and consider, first, the homogeneous system

$$\dot{x}_i = \sum_{j=1}^{n} a_{ij}(t) x_j, \qquad i = 1, 2, \ldots, n,$$

which we abbreviate to

$$\dot{\boldsymbol{x}} = \boldsymbol{A}(t)\boldsymbol{x}.$$

We assume that the determinant $|\, \boldsymbol{A}(t)\, |$ is not equal to zero.

It is clear that any combination of particular solutions $\boldsymbol{x}^{(1)}(t)$, $\boldsymbol{x}^{(2)}(t)$, ... is also a solution. Let us, then, choose n solutions $\boldsymbol{x}^{(1)}(t)$, $\boldsymbol{x}^{(2)}(t)$, ..., $\boldsymbol{x}^{(n)}(t)$ so that

$$\boldsymbol{x}^{(1)}(0) = \begin{pmatrix} 1 \\ 0 \\ \cdot \\ \cdot \\ \cdot \\ 0 \end{pmatrix}, \qquad \boldsymbol{x}^{(2)}(0) = \begin{pmatrix} 0 \\ 1 \\ \cdot \\ \cdot \\ \cdot \\ 0 \end{pmatrix},$$

$$\cdots\cdots, \qquad \boldsymbol{x}^{(n)}(0) = \begin{pmatrix} 0 \\ \cdot \\ \cdot \\ \cdot \\ 0 \\ 1 \end{pmatrix}.$$

These are obviously linearly independent. Further, let $\boldsymbol{x}(t)$ be a solution with the initial conditions

$$\boldsymbol{x}(0) = \mathring{\boldsymbol{x}}.$$

The solution

$$\sum_{j=1}^{n} \mathring{x}_j \boldsymbol{x}^{(j)}(t)$$

also satisfies these initial conditions. Therefore, by the uniqueness theorem of II § 2.2,

$$\boldsymbol{x}(t) = \sum_{j=1}^{n} \mathring{x}_j \boldsymbol{x}^{(j)}(t).$$

Every solution of the system of equations is, therefore, a linear combination of the $\boldsymbol{x}^{(i)}(t)$. We can obviously substitute for them any n linearly independent particular solutions.

Let us write

$$\boldsymbol{X}(t) = (\boldsymbol{x}^{(1)}, \boldsymbol{x}^{(2)}, \ldots, \boldsymbol{x}^{(n)}) = \begin{pmatrix} x_1^{(1)} & x_1^{(2)} & \ldots & x_1^{(n)} \\ x_2^{(1)} & x_2^{(2)} & \ldots & x_2^{(n)} \\ \cdot & \cdot & & \cdot \\ \cdot & \cdot & & \cdot \\ \cdot & \cdot & & \cdot \\ x_n^{(1)} & x_n^{(2)} & \ldots & x_n^{(n)} \end{pmatrix},$$

then

$$\boldsymbol{x}(t) = \boldsymbol{X}(t)\mathring{\boldsymbol{x}}.$$

If we write the system of differential equations in the form

$$\boldsymbol{x} = \mathring{\boldsymbol{x}} + \int_0^t \boldsymbol{A}\boldsymbol{x}\, \mathrm{d}\tau,$$

then, by the method of successive approximations of II § 2.2, the formal general solution can be readily obtained. Thus:

$$\begin{aligned}
\overset{1}{\boldsymbol{x}} &= \mathring{\boldsymbol{x}} + \int_0^t \boldsymbol{A}\mathring{\boldsymbol{x}}\, \mathrm{d}\tau \\
&= \mathring{\boldsymbol{x}} + \boldsymbol{A}_1\mathring{\boldsymbol{x}}, \qquad\qquad \boldsymbol{A}_1 = \int_0^t \boldsymbol{A}\, \mathrm{d}\tau, \\
\overset{2}{\boldsymbol{x}} &= \mathring{\boldsymbol{x}} + \int_0^t \boldsymbol{A}\overset{1}{\boldsymbol{x}}\, \mathrm{d}\tau \\
&= \mathring{\boldsymbol{x}} + \boldsymbol{A}_1\mathring{\boldsymbol{x}} + \boldsymbol{A}_2\mathring{\boldsymbol{x}}, \qquad \boldsymbol{A}_2 = \int_0^t \boldsymbol{A}\boldsymbol{A}_1\, \mathrm{d}\tau, \\
&\cdots\cdots\cdots \\
\overset{k}{\boldsymbol{x}} &= (\boldsymbol{A}_0 + \boldsymbol{A}_1 + \boldsymbol{A}_2 + \ldots \boldsymbol{A}_k)\mathring{\boldsymbol{x}},
\end{aligned}$$

where

$$\boldsymbol{A}_0 = \boldsymbol{U} = \begin{pmatrix} 1 & 0 & \ldots & 0 \\ 0 & 1 & & 0 \\ \cdot & \cdot & & \cdot \\ \cdot & \cdot & & \cdot \\ \cdot & \cdot & & \cdot \\ 0 & 0 & \ldots & 1 \end{pmatrix}$$

is the unit matrix, and

$$\boldsymbol{A}_{\lambda+1} = \int_0^t \boldsymbol{A}\boldsymbol{A}_\lambda\, \mathrm{d}\tau, \qquad \lambda = 0, 1, 2, \ldots.$$

If $\lambda \to \infty$, $\overset{\lambda}{\boldsymbol{x}}$ tends towards the solution $\boldsymbol{x}(t)$, which here appears in the form of "Neumann's series". It converges for all t if the elements of $\boldsymbol{A}$ are bounded. The matrix $\boldsymbol{X}(t)$, therefore, is given by,

$$\boldsymbol{X}(t) = \sum_{\lambda=0}^{\infty} \boldsymbol{A}_\lambda(t).$$

6. We now solve the inhomogeneous system,

$$\dot{\boldsymbol{x}} = \boldsymbol{A}(t)\boldsymbol{x} + \boldsymbol{b}(t),$$

by Lagrange's method of variation of the parameters. (Section 2.)

Let $\boldsymbol{X}\mathring{\boldsymbol{x}}$ be the general solution of the homogeneous system; then, for the inhomogeneous system we try the form $\boldsymbol{Xy}$ where

$$\boldsymbol{y} = \boldsymbol{y}(t),$$

so that

$$(\boldsymbol{Xy})^{\cdot} = \boldsymbol{A}(\boldsymbol{Xy}) + \boldsymbol{b}.$$

Therefore,

$$\boldsymbol{X}\dot{\boldsymbol{y}} = \boldsymbol{b},$$

$$\boldsymbol{y} = \mathring{\boldsymbol{x}} + \int_0^t \boldsymbol{X}^{-1}(\tau)\boldsymbol{b}(\tau)\,\mathrm{d}\tau,$$

and finally,

$$\boldsymbol{x} = \boldsymbol{X}\mathring{\boldsymbol{x}} + \boldsymbol{X}\int_0^t \boldsymbol{X}^{-1}\boldsymbol{b}\,\mathrm{d}\tau.$$

The general solution of the inhomogeneous system can, therefore, be represented as the sum of the general solution of the homogeneous system and a particular solution of the inhomogeneous system given by $\boldsymbol{X}\int_0^t \boldsymbol{X}^{-1}\boldsymbol{b}\,\mathrm{d}\tau$.

7. If the coefficient matrix A of the linear homogeneous system,

$$\dot{\boldsymbol{x}} = \boldsymbol{Ax},$$

is constant, the solution may be obtained by the following method. We assume a particular solution to be given in the form $\mathring{\boldsymbol{x}}e^{\lambda t}$, where λ denotes a not yet determined constant. If this expression is substituted in the system of equations, it gives, as a necessary condition for $\mathring{\boldsymbol{x}}$ and λ, the equation

$$(\boldsymbol{A} - \lambda \boldsymbol{U})\mathring{\boldsymbol{x}} = 0. \tag{1}$$

If this linear, homogeneous system of algebraic equations is to have a non-zero solution $\mathring{\boldsymbol{x}}$, then its determinant must vanish:

$$|\boldsymbol{A} - \lambda \boldsymbol{U}| = 0. \tag{2}$$

This is an equation of the nth degree in λ, the "characteristic" or "secular equation". In general it will have n distinct solutions, $\lambda_1, \lambda_2, \ldots, \lambda_n$. In astronomical applications we may assume that these roots are indeed distinct. If they should not be so at first, then we can always make them so by an arbitrarily small variation of the coefficients. Such variations are permitted, since the coefficients depend on empirical data, which are known with only limited accuracy.

Since the coefficients of the equation are real its roots are either real or pairs of complex conjugates. The same is true of the "eigenvectors," $\mathring{\boldsymbol{x}}$, belonging to them. We shall work, at first, with numbers that are possibly complex, and only later separate them into real and imaginary parts by means of Euler's theorem (§ 2.2).

To each "eigenvalue" λ_i there belongs an eigenvector $\mathring{\boldsymbol{x}}^{(i)}$. Since its magnitude is arbitrary, we choose $|\mathring{\boldsymbol{x}}^{(i)}|$ to be unity. The n eigenvectors are linearly independent if $|\boldsymbol{A}| \neq 0$, which was assumed at the beginning (§ 5). Therefore the general solution can be written:

$$\boldsymbol{x} = \sum_{i=1}^{n} \alpha_i \mathring{\boldsymbol{x}}^{(i)} \exp \lambda_i t, \tag{3}$$

where the α_i are (scalar) constants which are determined by the initial conditions

$$\mathring{\boldsymbol{x}} = \sum_{i=1}^{n} \alpha_i \mathring{\boldsymbol{x}}^{(i)}.$$

The right-hand side must (by the uniqueness theorem of II § 2.2) be real. The explicit real representation can be obtained by means of Euler's theorem (§ 2.2).

8. Systems of linear homogeneous equations with constant coefficients occur frequently in the investigation of stability by the "method of small oscillations". Suppose the system of equations

$$\dot{x}_i = f_i(\boldsymbol{x}) \qquad i - 1, 2, \ldots, n,$$

has the "stationary solution" $\boldsymbol{x} = \mathring{\boldsymbol{x}}$. We wish to know the behaviour of those solutions which are close to $\mathring{\boldsymbol{x}}$ at the time $t = 0$. The method of § 4 can be applied to this problem. It is more usual, however, to use a somewhat different method. The functions f_i are developed in power series of $(x_i - \mathring{x}_i)$, which are terminated at linear terms. Thus we obtain the Jacobi variational equation,

$$(x_i - \mathring{x}_i)^{\cdot} = \sum_{j=1}^{n} \frac{\partial f_i(\mathring{\boldsymbol{x}})}{\partial x_j} (x_j - \mathring{x}_j),$$

again, and then we proceed as in section 7. We hope that, from the solutions of the approximate equations, we shall be able to decide whether or not the solutions of the exact equations stay near to the stationary solution at all times.

We cannot obtain a definite answer to this problem in this way. It can be shown that the result of this method may be false by the following example. Let α be any non-zero constant, then the functions

$$\left.\begin{aligned} x &= \frac{\cos(t+\alpha)}{t+\alpha}, \\ y &= \frac{\sin(t+\alpha)}{t+\alpha} \end{aligned}\right\}$$

satisfy the system of differential equations

$$\left.\begin{aligned} \dot{x} &= -y - x\sqrt{(x^2+y^2)}, \\ \dot{y} &= x - y\sqrt{(x^2+y^2)}. \end{aligned}\right\}$$

A stationary solution is $x \equiv 0$, $y \equiv 0$. The corresponding Jacobi variational equations read

$$\left.\begin{aligned} \dot{x} &= -y, \\ \dot{y} &= x. \end{aligned}\right\}$$

They have the particular solution

$$\left.\begin{aligned} x &= \frac{1}{\alpha}\cos(t+\alpha), \\ y &= \frac{1}{\alpha}\sin(t+\alpha), \end{aligned}\right\}$$

which, at $t = 0$, assumes the same initial values $\left(\dfrac{\cos\alpha}{\alpha}, \dfrac{\sin\alpha}{\alpha}\right)$ as the solution given for the original system of equations. Let us now choose $\alpha < 0$ and $|\alpha| \gg 1$, so that the common starting point is in the neighbourhood of zero. Although the solution of the variational equations stays near to the stationary solution at all times, it is "unstable," since the magnitude of at least one of the rigorous solutions increases indefinitely as $t \rightarrow -\alpha$.

Naturally, for a sufficiently small time interval, the two solutions are arbitrarily close. In general, however, no more can be said. We will, therefore, use this method only occasionally, although we recognize that it has been useful in many cases, and although scarcely any more powerful general method is available.

9. In I § 5.4 we solved Kepler's equation

$$\psi = e \sin (M + \psi)$$

by successive approximations. The original zero order approximation was the constant zero. Later, however, we substituted for this another, better approximation, which we obtained by approximating the right-hand side with a linear function of the unknown ψ and solving for ψ. We can proceed in a similar way with differential equations. The system

$$\dot{x}_i = f_i(\boldsymbol{x}, t), \qquad i = 1, \ldots, n,$$

can be approximated to by the linear system

$$(x_i - \mathring{x}_i)^{\cdot} = f_i(\mathring{\boldsymbol{x}}, t) + \sum_{j=1}^{n} \frac{\partial f_i(\mathring{\boldsymbol{x}}, t)}{\partial \mathring{x}_j} (x_j - \mathring{x}_j),$$
$$i = 1, \ldots, n,$$

and the solution of this used as a zero order approximation for the original system.

If we now apply this method to the equations given by the variation of the constants (see §§ 1 and 2) to the case of the planetary system, then the functions $f_i(\boldsymbol{x}, t)$ and the coefficients $\partial f_i(\mathring{\boldsymbol{x}}, t)/\partial \mathring{x}_j$, of the homogeneous linear system of equations, are all periodic functions of the time provided that the Kepler motion corresponding to the given initial conditions is commensurable: otherwise they are almost periodic functions. In practice this distinction is without significance (see § 2.3), since, within the limits of observational accuracy, almost periodic functions are periodic. We consider, therefore, in the following, a homogeneous linear system of equations,

$$\dot{\boldsymbol{x}}(t) = \boldsymbol{A}(t)\boldsymbol{x}(t),$$

with a periodic coefficient matrix $\boldsymbol{A}(t) = \boldsymbol{A}(t + P)$, where P is a positive constant.

The solution may be characterized analytically in the following manner. On account of the periodicity, $\boldsymbol{x}(t + P)$ is a solution of the system if $\boldsymbol{x}(t)$ is. By section 5, this can be represented in the form

$$\boldsymbol{x}(t + P) = \boldsymbol{X}(t)\boldsymbol{x}(P)$$

or,

$$\boldsymbol{X}(t + P) = \boldsymbol{X}(t)\boldsymbol{X}(P). \tag{1}$$

We now search for the particular initial vectors $\mathring{\boldsymbol{x}}$ for which

$$\boldsymbol{x}(t + P) = \lambda \boldsymbol{x}(t)$$

and, therefore,

$$\boldsymbol{x}(t + P) = \lambda \boldsymbol{X}(t)\mathring{\boldsymbol{x}},$$

where λ is a suitable constant. For these $\mathring{\boldsymbol{x}}$,

$$\boldsymbol{X}(t)\boldsymbol{X}(P)\mathring{\boldsymbol{x}} = \lambda \boldsymbol{X}(t)\mathring{\boldsymbol{x}}, \qquad \text{(by (1))}$$

or

$$\boldsymbol{X}(t)\{\boldsymbol{X}(P) - \lambda \boldsymbol{U}\}\mathring{\boldsymbol{x}} = 0$$

for all t, and especially for $t = 0$, when $\boldsymbol{X}(0) = \boldsymbol{U}$. Therefore,

$$\{\boldsymbol{X}(P) - \lambda \boldsymbol{U}\}\mathring{\boldsymbol{x}} = 0 \tag{2}$$

is a necessary and sufficient condition for $\mathring{\boldsymbol{x}}$. If the linear homogeneous system of algebraic equations is to possess a non-zero solution, then its determinant must vanish:

$$| \boldsymbol{X}(P) - \lambda \boldsymbol{U} | = 0. \tag{3}$$

For the solutions of this equation, and the solutions $\mathring{\boldsymbol{x}}$ of the linear system belonging to them, similar considerations to those of § 7 hold. In particular we may assume the eigenvalues λ_i to be distinct and the eigenvectors $\mathring{\boldsymbol{x}}^{(i)}$ to be linearly independent.

Let, now, λ_1 be some particular one of the n roots, $\mathring{\boldsymbol{x}}^{(1)}$ the eigenvector belonging to it, and $\boldsymbol{x}^{(1)}(t)$ the particular solution for which

$$\boldsymbol{x}^{(1)}(0) = \mathring{\boldsymbol{x}}^{(1)}.$$

Then from

$$x_i^{(1)}(t + P) = \lambda_1 x_i^{(1)}(t), \qquad i = 1, 2, \ldots, n,$$

follows a linear inhomogeneous difference equation for $\ln x_i^{(1)}(t)$:

$$\ln x_i^{(1)}(t + P) - \ln x_i^{(1)}(t) = \ln \lambda_1 \tag{4}$$

(λ_1 is different from zero, since, otherwise, $| X(P) | = 0$, from which it would follow that $| X(0) | = 0$, whereas, in fact, $| X(0) | = | U | = 1$.) Obviously, the general solution of (4) is

$$\ln x_i^{(1)}(t) = \frac{\ln \lambda_1}{P} t + \ln y_i^{(1)}(t),$$

where $\ln y_i^{(1)}(t)$ is a periodic function with the period P. From this it follows that

$$\boldsymbol{x}^{(1)}(t) = \lambda_1^{t/P}\boldsymbol{y}^{(1)}(t)$$

or, if we write

$$\ln \lambda_1 = \mu_1 + \nu_1\sqrt{-1},$$

$$\boldsymbol{x}^{(1)}(t) = \exp\,(\mu_1/P)t.\,\left(\cos\frac{\nu_1}{P}t + \sin\frac{\nu_1}{P}t\sqrt{-1}\right)\boldsymbol{y}^{(1)}(t).$$

Thus, the solution $x^{(1)}(t)$ is, in general, not a periodic function of t at all, or, if it is, its period is different from that of the coefficients of the differential equations.

We construct the solution matrix $\boldsymbol{X}(t)$, for example, by the method of § 5, for the interval $0 \ldots P$, and then we solve equation (1)

$$\{\boldsymbol{X}(P) - \lambda_1\boldsymbol{U}\}\mathring{\boldsymbol{x}}^{(1)} = 0$$

for $\mathring{\boldsymbol{x}}^{(1)}$, and put $\boldsymbol{x}^{(1)} = \boldsymbol{X}\mathring{\boldsymbol{x}}^{(1)}$. The general solution then is (as in § 7)

$$\boldsymbol{x}(t) = \sum_{j=1}^{n} \alpha_j\boldsymbol{x}^{(j)}(t),$$

where the scalar constants α_j are determined by the initial conditions

$$\boldsymbol{x}(0) = \sum_{j=1}^{n} \alpha_j\boldsymbol{x}^{(j)}(0).$$

4. The perturbation equations of celestial mechanics

1. We shall now apply some of the methods just sketched in § 3 to the problems of celestial mechanics. (Historically, much of this work was developed solely for this purpose.) The different methods that will be worked out are distinguished by

(i) the choice of the coordinates or of the varying parameters;
(ii) the methods by which the basic equations for the required coordinates (or for the varying parameters) are obtained;
(iii) the methods by which these equations are integrated;
(iv) the choice of the zero order approximation.

All possible combinations are used, (without the different points of view always being clearly distinguished). The first point is of no great importance for us. We shall always (as hitherto) employ only Cartesian or polar coordinates; and, as variable parameters, we shall take only those quantities which appear as constants of integration in the solution of the two body problem. In this way we avoid the explicit theory of coordinate transformations (Hamilton's theory). For the derivation, (ii), and integration, (iii), of the equations, we give only the simplest but, just because of that, the most effective methods. As our zero order approximation, (iv), we take, in general, the Kepler motion.

2. Let

$$\left.\begin{aligned} \dot{\boldsymbol{r}} - \boldsymbol{v} &= 0, \\ \dot{\boldsymbol{v}} + \frac{G\mu}{r^2}\frac{\boldsymbol{r}}{r} &= \boldsymbol{s} \end{aligned}\right\}$$

be the equations of motion for, say, a planet, where $\boldsymbol{s}$ denotes some kind of disturbing force per unit mass. Further, let $\phi(\boldsymbol{r}, \boldsymbol{v}, t)$ be a first integral of the equations of motion when $\boldsymbol{s} \equiv 0$ so that, then,

$$\phi\{\boldsymbol{r}(t), \boldsymbol{v}(t), t\} \equiv C,$$

where C is a constant. If $\boldsymbol{s} \not\equiv 0$, then, by § 3.2 the perturbation equation for C is

$$\boxed{\dot{C} = \mathrm{grad}_{\boldsymbol{v}}\, \phi \,.\, \boldsymbol{s},} \qquad (1)$$

where $\mathrm{grad}_{\boldsymbol{v}}\, \phi$ is the vector

$$\begin{pmatrix} \dfrac{\partial \phi}{\partial v_x} \\[2mm] \dfrac{\partial \phi}{\partial v_y} \\[2mm] \dfrac{\partial \phi}{\partial v_z} \end{pmatrix}$$

and v_x, v_y, v_z are the components of the velocity vector $\boldsymbol{v}$. For the purposes of differentiation, these components are to be considered as independent variables. The "Poisson" equation (1) is the fundamental equation in our theory of perturbations.

3. We first apply the fundamental perturbation equation to the energy integral

$$h = \tfrac{1}{2}\boldsymbol{v}^2 - \frac{G\mu}{r}.$$

It gives at once

$$\dot{h} = \boldsymbol{v} \,.\, \boldsymbol{s}. \tag{1}$$

Now, by II § 3.7, the energy constant h is related to the semi-major axis, a, by

$$h = -\frac{G\mu}{2a}.$$

By differentiating this with respect to the time variable, and combining the result with (1) we obtain:

$$\dot{a} = 2\frac{a^2}{G\mu}\,\boldsymbol{v} \,.\, \boldsymbol{s}. \tag{2}$$

By I § 5.6, the vectors $\boldsymbol{r}$ and $\boldsymbol{v}$ are here to be understood as known functions of the elements and of the time.

4. For the rate of change of the integral of areas,

$$\boldsymbol{c} = \tfrac{1}{2}(\boldsymbol{r} \times \boldsymbol{v}),$$

with time, Poisson's method gives the differential equation

$$\dot{\boldsymbol{c}} = \tfrac{1}{2}(\boldsymbol{r} \times \boldsymbol{s}). \tag{1}$$

The vector $\boldsymbol{c}$ can now be expressed in terms of the elements Ω, ι, e (and a), by means of the formulae of II § 2.2. Alternatively, in reverse, the elements on the right-hand side can be expressed in terms of the vector $\boldsymbol{c}$. This second way is to be preferred, on account of the uncertainty in the elements Ω and e discussed in I § 6.5. The elements are then obtained from the vector $\boldsymbol{c}$ only at the end of the computations (if at all).

We explain the details. First we set up differential equations for e, Ω and ι. Let

$$\begin{pmatrix} \sin\Omega\sin\iota \\ -\cos\Omega\sin\iota \\ \cos\iota \end{pmatrix} \equiv \boldsymbol{k}, \tag{2}$$

so that, by II § 2.2

$$\boldsymbol{c} = c\boldsymbol{k}.$$

Therefore

$$\dot{\boldsymbol{c}} = \dot{c}\boldsymbol{k} + c\frac{\partial\boldsymbol{k}}{\partial\Omega}\dot{\Omega} + c\frac{\partial\boldsymbol{k}}{\partial\iota}\frac{\mathrm{d}\iota}{\mathrm{d}t}. \tag{3}$$

Now, as the reader may easily verify,

$$\left.\begin{array}{lll} \boldsymbol{k} \,.\, \boldsymbol{k} = 1, & \boldsymbol{k} \,.\, \dfrac{\partial \boldsymbol{k}}{\partial \Omega} = 0, & \boldsymbol{k} \,.\, \dfrac{\partial \boldsymbol{k}}{\partial \iota} = 0, \\ & \dfrac{\partial \boldsymbol{k}}{\partial \Omega} \,.\, \dfrac{\partial \boldsymbol{k}}{\partial \Omega} = \sin^2 \iota, & \dfrac{\partial \boldsymbol{k}}{\partial \Omega} \,.\, \dfrac{\partial \boldsymbol{k}}{\partial \iota} = 0, \\ & & \dfrac{\partial \boldsymbol{k}}{\partial \iota} \,.\, \dfrac{\partial \boldsymbol{k}}{\partial \iota} = 1. \end{array}\right\}$$

It follows, therefore, by introducing (3) in (1) and taking the scalar products with $\boldsymbol{k}$, $\dfrac{\partial \boldsymbol{k}}{\partial \Omega}$, $\dfrac{\partial \boldsymbol{k}}{\partial \iota}$ in turn, that

$$\left.\begin{aligned} \dot{c} &= \tfrac{1}{2}[\boldsymbol{r}, \boldsymbol{s}, \boldsymbol{k}], \\ c \sin^2 \iota \dot{\Omega} &= \frac{1}{2}\left[\boldsymbol{r}, \boldsymbol{s}, \frac{\partial \boldsymbol{k}}{\partial \Omega}\right], \\ c \frac{\mathrm{d}\iota}{\mathrm{d}t} &= \frac{1}{2}\left[\boldsymbol{r}, \boldsymbol{s}, \frac{\partial \boldsymbol{k}}{\partial \iota}\right]. \end{aligned}\right\} \tag{4}$$

If we differentiate the relation

$$c = \tfrac{1}{2}\sqrt{[G\mu a(1 - e^2)]}, \tag{5}$$

(II § 3.4(6)), with respect to t, we can eliminate $\dot{c}$ by the first equation of (4) to obtain a differential equation for e. It is simpler to regard e as a variable dependent on c, and so to substitute for e in all formulae by the expression

$$e = \left[1 - \frac{4c^2}{G\mu a}\right]^{1/2}, \tag{6}$$

which follows from (5).

If we wish to use the equations (4) as they stand, and so to consider c as a dependent variable, then we must express e, ι, Ω, in the formula I § 5.6(5), in terms of the components of $\boldsymbol{c}$. Although this is readily done, we postpone the transformation, since it is essentially simplified by results to be developed in the next section.

5. The existence of the integral of areas, $\tfrac{1}{2}(\boldsymbol{r} \times \boldsymbol{v})$, expresses the fact that an undisturbed planet moves in a plane fixed in space. Similarly, the fact that the direction of perihelion is fixed in space must also be able to be expressed as an integral. The perihelion direction can also be characterized by a unit

vector, and so two corresponding integrals are obtained. They cannot both be independent of the integral of areas however, since the direction of perihelion must lie in the orbital plane and, therefore, be perpendicular to the vector $(\boldsymbol{r} \times \boldsymbol{v})$. We can take the first integral we need from the solution of the two body problem given in II § 3. It is more convenient, however, to derive it afresh by the following method.

We take the vector product of the equation of motion of the two body system,

$$\dot{\boldsymbol{v}} = -\frac{G\mu}{r^2}\frac{\boldsymbol{r}}{r}$$

with

$$\boldsymbol{c} = \tfrac{1}{2}(\boldsymbol{r} \times \boldsymbol{v}).$$

That gives

$$\boldsymbol{c} \times \dot{\boldsymbol{v}} = -\frac{1}{2}\frac{G\mu}{r^2}(\boldsymbol{r} \times \boldsymbol{v}) \times \frac{\boldsymbol{r}}{r}.$$

Now, for any three vectors $\boldsymbol{a}$, $\boldsymbol{b}$, $\boldsymbol{c}$ the "Lagrange identity",

$$(\boldsymbol{a} \times \boldsymbol{b}) \times \boldsymbol{c} = (\boldsymbol{a} \,.\, \boldsymbol{c})\boldsymbol{b} - (\boldsymbol{b} \,.\, \boldsymbol{c})\boldsymbol{a}, \tag{1}$$

holds. The right-hand side therefore becomes

$$-\frac{1}{2}\frac{G\mu}{r^2}\frac{(\boldsymbol{r} \,.\, \boldsymbol{r})\boldsymbol{v} - (\boldsymbol{v} \,.\, \boldsymbol{r})\boldsymbol{r}}{r}$$

which, since $\boldsymbol{r} \,.\, \boldsymbol{r} = r^2$ and $\boldsymbol{v} \,.\, \boldsymbol{r} = \dot{\boldsymbol{r}} \,.\, \boldsymbol{r} = \dot{r}r$, may be written

$$-\frac{1}{2}\frac{G\mu}{r^2}(r\dot{\boldsymbol{r}} - \dot{r}\boldsymbol{r})$$
$$= -\tfrac{1}{2}G\mu\frac{\mathrm{d}}{\mathrm{d}t}\left(\frac{\boldsymbol{r}}{r}\right).$$

So we obtain

$$\frac{\mathrm{d}}{\mathrm{d}t}(\boldsymbol{c} \times \boldsymbol{v}) + \tfrac{1}{2}G\mu\frac{\mathrm{d}}{\mathrm{d}t}\left(\frac{\boldsymbol{r}}{r}\right) = 0,$$

$$\boldsymbol{c} \times \boldsymbol{v} + \tfrac{1}{2}G\mu\frac{\boldsymbol{r}}{r} = -\tfrac{1}{2}\boldsymbol{p} = \text{const.},$$

that is

$$(\boldsymbol{r} \times \boldsymbol{v}) \times \boldsymbol{v} + G\mu\frac{\boldsymbol{r}}{r} = -\boldsymbol{p} \tag{2}$$

so that (again by the identity (1)),

$$-(\boldsymbol{r} \,.\, \boldsymbol{v})\boldsymbol{v} + (\boldsymbol{v} \,.\, \boldsymbol{v})\boldsymbol{r} - G\mu\frac{\boldsymbol{r}}{r} = \boldsymbol{p}. \tag{3}$$

If, therefore, $\boldsymbol{r}$ and $\boldsymbol{v}$ are taken as independent variables, the left-hand side represents three first integrals.

If we substitute for $\boldsymbol{r}$ and $\boldsymbol{v}$ their values, $r_p\boldsymbol{i}$ and $\boldsymbol{v}_p$, at the moment of the planet's passage through its perihelion, then $\boldsymbol{i} \,.\, \boldsymbol{v}_p = 0$ and

$$\boldsymbol{p} = (\boldsymbol{v}_p^2 r_p - G\mu)\boldsymbol{i}$$

—the vector $\boldsymbol{p}$ lies along the line of apsides, (I § 4.2). To show that $\boldsymbol{p}$ actually points towards the perihelion (and not in the opposite direction), we transform the last expression in the following way:

$$\boldsymbol{p} = \left(\frac{\boldsymbol{v}_p^2 r_p^2}{r_p} - G\mu\right)\boldsymbol{i}$$

$$= \left(\frac{4c^2}{r_p} - G\mu\right)\boldsymbol{i},$$

since

$$c = \tfrac{1}{2} r_p \,|\, \boldsymbol{v}_p \,|.$$

We also have

$$c^2 = \tfrac{1}{4}(1 - e^2)aG\mu,$$

(§ 3.2) and

$$r_p = a(1 - e),$$

so that

$$\boldsymbol{p} = G\mu e\boldsymbol{i}. \tag{4}$$

From this we deduce that $\boldsymbol{p}$ is directed towards the perihelion—as we asserted.

To express $\boldsymbol{p}$ completely in terms of the orbital elements, we must represent the perihelion direction $\boldsymbol{i}$ by equation I § 5.6(6). Then we have

$$\boldsymbol{p} = G\mu e \begin{pmatrix} \cos\omega\cos\Omega - \sin\omega\sin\Omega\cos\iota \\ \cos\omega\sin\Omega + \sin\omega\cos\Omega\cos\iota \\ \sin\omega \qquad \sin\iota \end{pmatrix}. \tag{5}$$

6. The equation for $\dot{\boldsymbol{p}}$ can readily be obtained by applying 2(1) to 5(2) or 5(3). This gives

$$\dot{\boldsymbol{p}} = (\boldsymbol{v} \times \boldsymbol{r}) \times \boldsymbol{s} + (\boldsymbol{s} \times \boldsymbol{r}) \times \boldsymbol{v}, \tag{1}$$

or

$$\dot{\boldsymbol{p}} = 2(\boldsymbol{v} \,.\, \boldsymbol{s})\boldsymbol{r} - (\boldsymbol{r} \,.\, \boldsymbol{s})\boldsymbol{v} - (\boldsymbol{v} \,.\, \boldsymbol{r})\boldsymbol{s}. \tag{2}$$

As in section 4, we may choose which of the unknown functions we shall determine first. If we favour the conventional elements, it is convenient to start with the functions e, ι, Ω, ω, or, rather, the functions ι, Ω, and

$$\left.\begin{aligned} C &= e \cos \omega, \\ S &= e \sin \omega. \end{aligned}\right\} \tag{3}$$

We have, therefore, to replace the first equation for $\dot{e}$, 4(3), by one each for $\dot{C}$ and $\dot{S}$. These are obtained by introducing the expressions (3) into 5(5), solving for C and S, and then differentiating with respect to the time. As we shall make no use of the resulting equation, we do not carry this out in detail.

If the components of the moment of momentum are to be retained as independent variables in the perturbation equation, then a suitable variable must be found which, together with the other variables, specifies the direction of perihelion. Such a variable, for example, is C. It is clearly possible to rewrite the equations for $\dot{C}$ and $\boldsymbol{r}$ (I § 5.6) in the new variables. Admittedly, the right-hand sides of these equations then become rather complicated, while that for $\boldsymbol{c}$ is unaltered, and, therefore, remains relatively simple. Again, we do not give these equations explicitly.

A thorough simplification of the differential equation is obtained if $\boldsymbol{c}$ and $\boldsymbol{p}$ are kept as dependent variables. Then, of course, we have two more unknown functions; the ("integral") relations already known between $\boldsymbol{c}$ and $\boldsymbol{p}$, for example the relation

$$\boldsymbol{c} \cdot \boldsymbol{p} = 0,$$

are not used in this case. These additional relations provide controls for the numerical calculations, (especially if we are satisfied with the first approximation). Moreover, in using the equations of motion themselves (with the coordinate origin in the Sun) we have tacitly renounced the use even of the integrals of energy and of moment of momentum. Their consideration would reduce the order of the system of differential equations, but would make the right-hand sides very involved—so the present procedure is only one step more in this direction.

We can use the equations for $\dot{\boldsymbol{c}}$ and $\dot{\boldsymbol{p}}$ at once, if we express the coefficients of equation I § 5.6(5),

$$\frac{\boldsymbol{r}}{a} = \boldsymbol{i}(\cos E - e) + \boldsymbol{j}\sqrt{(1 - e^2)} \sin E, \tag{4}$$

and of the corresponding equation for the velocity vector $\boldsymbol{v} = \dot{\boldsymbol{r}}$, in terms of $\boldsymbol{c}$ and $\boldsymbol{p}$. This can easily be done, since

$$\boldsymbol{i} = \frac{\boldsymbol{p}}{G\mu e} \quad \text{(by 5(5))}, \tag{5}$$

and

$$\boldsymbol{j} = \boldsymbol{k} \times \boldsymbol{i} = \frac{\boldsymbol{c}}{c} \times \frac{\boldsymbol{p}}{G\mu e}, \tag{6}$$

where, again $c = |\, \boldsymbol{c} \,|$, and

$$e = \left(1 - \frac{4c^2}{G\mu a}\right)^{1/2}$$

(by 4(6)).

We obtain the velocity vector $\boldsymbol{v}$ by differentiating (4) with respect to t. At first the derivative $\dot{E}$ appears in the result. Differentiating Kepler's equation,

$$E - e \sin E = \frac{\sqrt{(G\mu)}}{a^{3/2}} (t - T),$$

with respect to t, we find that

$$\dot{E} = \frac{\sqrt{(G\mu)}}{a^{3/2}(1 - e \cos E)},$$

so that, finally,

$$\boldsymbol{v} = \frac{\sqrt{(G\mu)}}{\sqrt{a}\,.\,(1 - e \cos E)} \left\{ -\frac{\boldsymbol{p}}{G\mu e} \sin E + \right.$$

$$\left. + (1 - e^2)^{1/2} \left(\frac{\boldsymbol{c}}{c} \times \frac{\boldsymbol{p}}{G\mu e} \right) \cos E \right\}. \tag{7}$$

Thus is the required transformation of variables completely given.

7. We can obtain a differential equation for the disturbed time of perihelion from Kepler's equation

$$E - e \sin E = \frac{\sqrt{(G\mu)}}{a^{3/2}} (t - T),$$

or,

$$T = t - \frac{a^{3/2}}{\sqrt{(G\mu)}} (E - e \sin E). \tag{1}$$

9

Applying the fundamental equation 2(1) we have

$$\begin{aligned}\dot{T} &= -\frac{1}{\sqrt{(G\mu)}}\,\boldsymbol{s}\,.\,\mathrm{grad}_{\boldsymbol{v}}\,\{a^{3/2}(E - e\sin E)\}\\ &= -(G\mu)^{-1/2}\boldsymbol{s}\,.\,[\mathrm{grad}_{\boldsymbol{v}}\,(a^{3/2})(E - e\sin E) +\\ &\quad + a^{3/2}\,\mathrm{grad}_{\boldsymbol{v}}\,(E - e\sin E)]\\ &= -(G\mu)^{-1/2}\boldsymbol{s}[\tfrac{3}{2}a^{1/2}\,\mathrm{grad}_{\boldsymbol{v}}\,a\,.\,(E - e\sin E) +\\ &\quad + a^{3/2}(1 - e\cos E)\,\mathrm{grad}_{\boldsymbol{v}}\,E -\\ &\quad - a^{3/2}\sin E\,\mathrm{grad}_{\boldsymbol{v}}\,e],\end{aligned} \tag{2}$$

where a, e and E are here considered as functions of $\boldsymbol{r}$ and $\boldsymbol{v}$. Now, by § 3,

$$a = -\frac{1}{2}\frac{G\mu}{h},$$

so

$$\begin{aligned}\mathrm{grad}_{\boldsymbol{v}}\,a &= \frac{G\mu}{2h^2}\,\mathrm{grad}_{\boldsymbol{v}}\,h,\\ &= 2\,\frac{a^2}{G\mu}\,\boldsymbol{v}.\end{aligned} \tag{3}$$

We evaluate $\mathrm{grad}_{\boldsymbol{v}}\,E$ from the relation I § 5.6(2)

$$r = a(1 - e\cos E),$$

that is from

$$1 - e\cos E = \frac{r}{a} = -2r\,\frac{h}{G\mu}. \tag{4}$$

Then

$$e\sin E\,.\,\mathrm{grad}_{\boldsymbol{v}}\,E - \mathrm{grad}_{\boldsymbol{v}}\,e\,.\,\cos E = -\frac{2r}{G\mu}\,\boldsymbol{v}. \tag{7}$$

Thus the computation of $\mathrm{grad}_{\boldsymbol{v}}\,E$ is made to depend on that of $\mathrm{grad}_{\boldsymbol{v}}\,e$, which is also needed in equation (2). By 3(4) e is given by

$$\begin{aligned}e^2 &= 1 - 4\,\frac{c^2}{G\mu a}\\ &= 1 + 8\,\frac{c^2 h}{G^2\mu^2}\end{aligned} \tag{6}$$

where

$$c^2 = \tfrac{1}{4}(\boldsymbol{r} \times \boldsymbol{v})^2. \tag{7}$$

Therefore

$$e \operatorname{grad}_v e = \frac{4}{G^2\mu^2} [\operatorname{grad}_v (c^2) \,.\, h + c^2 \operatorname{grad}_v h]. \tag{8}$$

Now, for any four vectors $\boldsymbol{a}$, $\boldsymbol{b}$, $\boldsymbol{c}$, $\boldsymbol{d}$, we have the identity

$$(\boldsymbol{a} \times \boldsymbol{b})(\boldsymbol{c} \times \boldsymbol{d}) = (\boldsymbol{a} \,.\, \boldsymbol{c})(\boldsymbol{b} \,.\, \boldsymbol{d}) - (\boldsymbol{a} \,.\, \boldsymbol{d})(\boldsymbol{b} \,.\, \boldsymbol{c}),$$

so that

$$c^2 = \tfrac{1}{4}(\boldsymbol{r}^2)(\boldsymbol{v}^2) - \tfrac{1}{4}(\boldsymbol{r} \,.\, \boldsymbol{v})^2 \tag{9}$$

and

$$\operatorname{grad}_v (\boldsymbol{c}^2) = \tfrac{1}{2}(\boldsymbol{r}^2)\boldsymbol{v} - \tfrac{1}{2}(\boldsymbol{r} \,.\, \boldsymbol{v})\boldsymbol{r}. \tag{10}$$

So we obtain

$$e \operatorname{grad}_v e = \frac{1}{G^2\mu^2}\left[-\frac{G\mu}{a}\left\{(\boldsymbol{r}^2)\boldsymbol{v} - (\boldsymbol{r} \,.\, \boldsymbol{v})\boldsymbol{r}\right\} + c^2\boldsymbol{v}]\right]. \tag{11}$$

Setting the expressions (3), (5), (6) and (11) in (2), we finally obtain the differential equation for T, the only equation, with a more complicated right-hand side. Again we do not write this equation explicitly since we make no further use of it.

The working could be made somewhat more simple by taking

$$\frac{\sqrt{(G\mu)}}{a^{3/2}}\, T$$

in place of T as the disturbed element.

8. We summarize. We write the equations of motion of § 2.1(1) in the form

$$\ddot{\boldsymbol{r}} + G\mu \frac{\boldsymbol{r}}{|\,\boldsymbol{r}\,|^3} = \boldsymbol{s}, \quad \text{etc.}, \tag{1}$$

We assume the solution $\boldsymbol{r}(t)$, $\boldsymbol{v}(t)$ to have the form

$$\left.\begin{aligned} \boldsymbol{r} &= a\frac{\boldsymbol{p}}{G\mu e}(\cos E - e) + (1 - e^2)^{1/2} \times \\ &\qquad \times \left(\frac{\boldsymbol{c}}{c} \times \frac{\boldsymbol{p}}{G\mu e}\sin E)\right), \\ \boldsymbol{v} &= \frac{\sqrt{(G\mu)}}{\sqrt{a}(1 - e\cos E)}\left\{-\frac{\boldsymbol{p}}{G\mu e}\sin E + \right. \\ &\qquad \left. + (1 - e^2)^{1/2}\left(\frac{\boldsymbol{c}}{c} \times \frac{\boldsymbol{p}}{G\mu e}\cos E\right\}\right. \end{aligned}\right\} \tag{2}$$

where

$$e = \left[1 - \frac{4c^2}{G\mu a}\right]^{1/2} \tag{3}$$

and

$$E = \frac{G^{1/2}\mu^{1/2}}{a^{3/2}}(t - T) + e \sin E. \tag{4}$$

Then the "elements" a, $\boldsymbol{c}$, $\boldsymbol{p}$, and T satisfy the eight differential equations of the first order:

$$\left.\begin{aligned} \dot{a} &= \frac{2}{G\mu} a^2 \boldsymbol{v} \cdot \boldsymbol{s}, \\ \dot{\boldsymbol{c}} &= \tfrac{1}{2}\boldsymbol{r} \times \boldsymbol{s}, \\ \dot{\boldsymbol{p}} &= -(\boldsymbol{r} \cdot \boldsymbol{s})\boldsymbol{v} + 2(\boldsymbol{v} \cdot \boldsymbol{s})\boldsymbol{r} - (\boldsymbol{r} \cdot \boldsymbol{v})\boldsymbol{s}, \end{aligned}\right\} \tag{5}$$

and the equation for $\dot{T}$ given by 7(2), where

$$\left.\begin{aligned} \boldsymbol{c} &= \tfrac{1}{2}(\boldsymbol{r} \times \boldsymbol{v}), \\ \boldsymbol{p} &= (\boldsymbol{v} \cdot \boldsymbol{v})\boldsymbol{r} - (\boldsymbol{r} \cdot \boldsymbol{v})\boldsymbol{v} - G\mu \frac{\boldsymbol{r}}{|\boldsymbol{r}|}. \end{aligned}\right\} \tag{6}$$

In equations (5), $\boldsymbol{r}$ and $\boldsymbol{v}$ may be eliminated on the right-hand sides by means of equations (2). We shall not do so here. It is also possible to set up a system of only six first order equations for the six conventional elements, or for some suitable combinations of them. This is the more usual way. The right-hand sides, however, would then be considerably more complicated, so that the way chosen here seems the more advantageous. If there are many bodies, each disturbing the others, then for each body a system of eight (or six) similar equations must be set up.

5. Perturbations in the motions of the planets

We now apply the results of § 4 to the case of two planets disturbing each other. In the general perturbation formula § 4.2(1)

$$\dot{C} = \boldsymbol{s}\, \mathrm{grad}_{\boldsymbol{v}}\, \phi \tag{1}$$

we write

$$\boldsymbol{s} = Gm'\left(\frac{\boldsymbol{r}' - \boldsymbol{r}}{|\boldsymbol{r}' - \boldsymbol{r}|^3} - \frac{\boldsymbol{r}'}{r'^3}\right). \tag{2}$$

Primed symbols refer to the disturbing and unprimed to the disturbed planet.

Since the disturbing masses are small, so also are the right-hand sides of the perturbation equations and the method of successive approximations (II § 2) is the obvious method for their integration. For the zero order approximation we take the constant initial values of the elements. If the succession of iterations is stopped at the first step, and in many cases this yields sufficient accuracy, then the system of equations uncouples itself—each equation can be integrated by itself independently of the other. In this case it is particularly useful to keep the vectors $\boldsymbol{c}$ and $\boldsymbol{p}$ as elements.

The Fourier development of this first equation can be found by the method of § 2.2. Let $\psi(E, E')$ denote the right-hand side of equation (1) after $\boldsymbol{r}$, $\boldsymbol{v}$, $\boldsymbol{r}'$ and $\boldsymbol{v}'$ have been introduced into it as functions of the elements and of the eccentric anomalies E and E'. Then, as in I § 5.3 and in III § 2.2

$$\psi(E, E') = \sum_{\nu=-\infty}^{+\infty} \sum_{\nu'=-\infty}^{+\infty} C_{\nu\nu'} \exp\left[-(\nu M + \nu' M')\sqrt{-1}\right], \tag{3}$$

where

$$C_{\nu\nu'} = \frac{1}{4\pi^2} \int_0^{2\pi}\int_0^{2\pi} \psi(E, E') \exp\left[\sqrt{-1}(\nu E + \nu' E' - \nu e \sin E - \right.$$
$$\left. - \nu' e' \sin E')\right] . (1 - e \cos E)(1 - e' \cos E')\, \mathrm{d}E\, \mathrm{d}E'. \tag{4}$$

All that has been said in § 2.2 holds for the evaluation of the $C_{\nu\nu'}$.

For "commensurability", small divisors appear in the series for the first integral ϕ, (see equation (1)) although only in the first power, since each series (3) will only be integrated once, while in the series for the position vectors these divisors appear squared (see § 2.2). These commensurabilities have been related with the gaps and accumulations in the distribution of the semi-axes of the minor planets (that is the commensurabilities with the motion of Jupiter). A relation is also supposed to exist between the partitions in the rings of Saturn and the periods of its moons. A satisfactory treatment of this problem from the point of view of celestial mechanics, however, seems still to be lacking.

It can be proved that in the Fourier development of the semi-major axis the secular term, C_{00}, vanishes. From this it has been attempted to draw conclusions about the "stability" of the planetary system. (The word "stability" has a different meaning here than in § 3.8. It means that all position vectors are uniformly bounded at all times.) If the system is stable, then the eccentricity e and the inclination ι can also perform only small oscillations.

All these arguments are not cogent mathematically, however, since they are concerned only with approximations even the best of which can be guaranteed only over a restricted time interval. Thus, at least in their present form, these reasonings have no relevance to the real planetary system, the mechanical relationships of which are only approximately described by the equations of motion. (For example the gravitational influence of the fixed stars has been neglected.) Even if the accuracy of the approximation of the forces is very great, nevertheless between the observed and computed positions of the planet arbitrarily large deviations can, in principle, appear after a sufficiently long time. For these reasons, we shall not consider these stability investigations any further. The significance of the "perturbation equations" lies only in their application to the observations. General conclusions about the forms of the motion cannot be drawn from them (see § 2.2).

Power series in the small masses can be used instead of successive approximations. The first of the successive approximations is identical with the linear term of the series. There is no such relation, however, between the approximations and terms of higher order. The perturbation equations may also be linearized and the methods of §§ 3.8 and 3.9 applied. The mixed analytic–numerical method just outlined seems to us, however, to be the most suitable.

6. Perturbations in the motion of the Moon

1. As was remarked in § 5, the principle difficulty in the analytical representation of perturbations is caused by the denominator $|\,\boldsymbol{r}_2 - \boldsymbol{r}_1\,|^3$. The perturbations of the satellites by the Sun may be represented in this way relatively simply, if only a moderate approximation is required. For the Moon this is not sufficient, since its motion, relative to the Earth, can be

very accurately observed. More exact results can again be obtained more simply by numerical computation of the Fourier coefficients. The disturbed Kepler motion, as obtained from the analytical approximation, may then be used instead of the undisturbed Kepler motion, on the right-hand side of equations (4) of § 5. This solution can be improved numerically, to an arbitrary degree of accuracy, by means of successive approximations. The analytical approximation, therefore, serves to make the principles clear and to provide an estimate of the order of magnitude of the perturbations, while numerical calculation provides the accurate predictions of the motion.

We shall now develop a simple analytical approximation. Let the coordinate origin be in the centre of the Earth. We use the notation of § 5, without the prime for the Moon, and with the prime for the Sun. The empirical value for the semi-major axis of the orbit of the Moon is

$$a = 2{\cdot}57 \times 10^{-3}. \tag{1}$$

Therefore we approximate the disturbing force

$$\boldsymbol{s} = G\left\{\frac{\boldsymbol{r}' - \boldsymbol{r}}{|\,\boldsymbol{r}' - \boldsymbol{r}\,|^3} - \frac{\boldsymbol{r}'}{r'^3}\right\} \tag{2}$$

in the following way:

$$\begin{aligned}
\boldsymbol{s} &= G\left[(\boldsymbol{r}' - \boldsymbol{r})\{(\boldsymbol{r}' - \boldsymbol{r})^2\}^{-3/2} - \frac{\boldsymbol{r}'}{r'^3}\right] \\
&= G\left[\frac{\boldsymbol{r}' - \boldsymbol{r}}{r'^3}\left\{1 - 2\frac{\boldsymbol{r}'}{r'}\,\boldsymbol{r} + O(a^2)\right\}^{-3/2} - \frac{\boldsymbol{r}'}{r'^3}\right] \\
&= G\left[\frac{\boldsymbol{r}' - \boldsymbol{r}}{r'^3}\left\{1 + 3\frac{\boldsymbol{r}'}{r'}\,\boldsymbol{r} + O(a^2)\right\} - \frac{\boldsymbol{r}'}{r'^3}\right] \\
&= G\left[\frac{\boldsymbol{r}'}{r'^3}\left\{3\frac{\boldsymbol{r}'}{r'}\,\boldsymbol{r} + O(a^2)\right\} - \frac{\boldsymbol{r}}{r'^3}\{1 + O(a)\}\right],
\end{aligned}$$

and, finally,

$$\boldsymbol{s} = G\frac{a}{r'^3}\left\{\left[3\frac{\boldsymbol{r}'}{r'}\frac{\boldsymbol{r}}{a} + O(a)\right]\boldsymbol{r}' - [1 + O(a)]\frac{\boldsymbol{r}}{a}\right\}. \tag{3}$$

Since the eccentricities, e and e', of the orbits of the Moon and Sun are small, we write

$$\boldsymbol{s} = Ga\left\{\left[3\boldsymbol{r}'\frac{\boldsymbol{r}}{a} + O(a)\right]\boldsymbol{r}' - \frac{\boldsymbol{r}}{a}\right\}\varepsilon, \tag{4}$$

where $\boldsymbol{r}$ and $\boldsymbol{r}'$ are now the position vectors of the Moon and Sun computed with $e = 0$, $e' = 0$, and

$$\varepsilon = 1 + O(e') + O(e) + O(a).$$

Since the coefficients of the quantities $O(\ldots)$ are of the order of magnitude of unity, the relative error in $|\boldsymbol{s}|$ is of the order of $(e' + e + a)$. The empirical values of e' and e are

$$\left.\begin{aligned} e' &= 0{\cdot}0168, \\ e &= 0{\cdot}0549. \end{aligned}\right\} \tag{5}$$

The possible error, therefore, is of the order of 7%, and the greater part, 5%, arises from the neglect of the eccentricity of the orbit of the Moon.

In what follows, we derive the perturbations of the semi-axis a, and of the vector $\boldsymbol{c}$ (i.e. of the longitude of the node, the orbital inclination and the eccentricity) under the simplifying assumption that the undisturbed motion is circular. Since we have already estimated the order of magnitude of the possible error we shall, for the sake of simplicity, omit the error terms in the equations.

2. First we derive the perturbing force $\boldsymbol{s}$ as a function of the time. Assuming that the undisturbed motion is circular, we can write

$$\left.\begin{aligned} \frac{\boldsymbol{r}}{a} &= \boldsymbol{i} \cos M + \boldsymbol{j} \sin M, \\ \boldsymbol{r}' &= \boldsymbol{i}' \cos M' + \boldsymbol{j}' \sin M'. \end{aligned}\right\} \tag{1}$$

Now we set

$$\boldsymbol{i}' = \boldsymbol{i}, \tag{2}$$

so that both these vectors lie along the nodal line of the orbital plane of the Moon (that is along the intersection of the orbital plane of the Moon with that of the Sun.) This is obviously possible for circular orbits—and only for them. From now on M and M' are to be understood to be the angles between $\boldsymbol{i}$ and the radius vectors of the Moon and Sun respectively. (This is different from their usual meaning.) We write, therefore,

$$\left.\begin{aligned} M &= \frac{2\pi}{P} t + \delta, \\ M' &= 2\pi t + \delta', \end{aligned}\right\} \tag{3}$$

where δ and δ' are two suitable constants.

Now, on account of (2),

$$\boldsymbol{j}' \cdot \boldsymbol{j} = \cos \iota = 1 + O(\iota^2),$$

$$\boldsymbol{j}' \cdot \boldsymbol{k} = -\cos\left(\frac{\pi}{2} - \iota\right) = -\iota + O(\iota^3),$$

therefore,

$$\boldsymbol{j}' = \boldsymbol{j} - \iota\boldsymbol{k} + O(\iota^2). \tag{4}$$

Since observation gives $\iota = 5^\circ{\cdot}1 \approx 1/11$, and since the error term in (4) has a coefficient of the order of unity, $O(\iota^2)$ is of the order of 1%. The total uncertainty (hitherto 7%) must be increased to 8% to allow for this new approximation. We shall also omit the term $O(\iota^2)$ in the rest of the working.

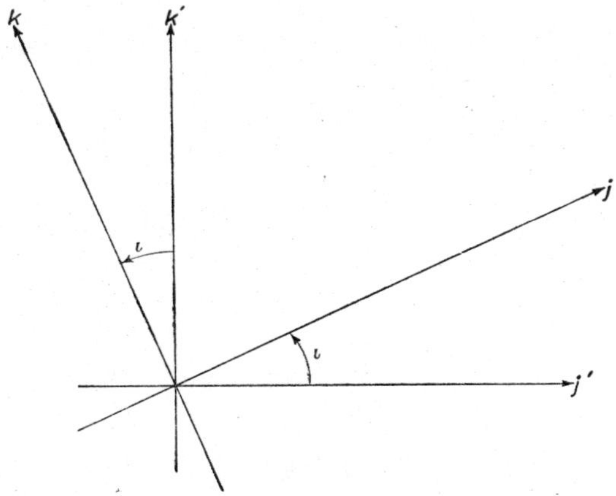

Fig. 3

From (1), (2) and (3), it follows that

$$\left.\begin{aligned} \frac{\boldsymbol{r}}{a} &= \boldsymbol{i}\cos M + \boldsymbol{j}\sin M, \\ \boldsymbol{r}' &= \boldsymbol{i}\cos M' + \boldsymbol{j}\sin M' - \boldsymbol{k}\iota\sin M', \end{aligned}\right\} \tag{5}$$

and therefore

$$\boldsymbol{r}'\frac{\boldsymbol{r}}{a} = \cos(M - M'),$$

and, after some elementary transformations, by means of the trigonometrical addition theorems, we obtain, finally,

$$\begin{aligned} \boldsymbol{s} = \tfrac{1}{2}Ga[\boldsymbol{i}\{\cos M + 3\cos(M - 2M')\} + \\ + \boldsymbol{j}\{\sin M - 3\sin(M - 2M')\} - \\ - \boldsymbol{k} \,.\, 3\iota\{\sin M - \sin(M - 2M')\}]. \end{aligned} \tag{6}$$

3. To compute the perturbations of the semi-major axis by the formula § 4.2(2)

$$\dot{a} = 2\frac{a^2}{G\mu}\boldsymbol{v} \,.\, \boldsymbol{s}, \tag{1}$$

we use the approximation

$$\boldsymbol{v} = \left(\frac{G\mu}{a}\right)^{1/2}[-\boldsymbol{i}\sin M + \boldsymbol{j}\cos M]$$

for the velocity vector $\boldsymbol{v}$. (See § 4.7(2).) Together with 2(5) this gives

$$\dot{a} = -3\frac{\sqrt{G}a^{5/2}}{\sqrt{\mu}}\sin 2(M - M'). \tag{2}$$

We eliminate μ, the total mass of the Earth and the Moon by means of the formula II § 3.7(7)

$$a^3 = \mu P^2.$$

Here P is the sidereal period of the Moon. We have set $G = 4\pi^2$, (cf. II § 1.6). Thus we obtain

$$\dot{a} = -6\pi a P\sin 2(M - M'). \tag{3}$$

On the right-hand side we now substitute for a and P the constant initial values $\mathring{a}$ and $\mathring{P}$ and integrate with respect to t. By 2(3):

$$M = \frac{2\pi}{P}t + \delta, \qquad M' = 2\pi t + \delta',$$

we have

$$a - \mathring{a} = \tfrac{3}{2}\mathring{a}\frac{\mathring{P}}{\dfrac{1}{\mathring{P}} - 1}[\cos 2(M - M') - \cos 2(\delta - \delta')]. \tag{4}$$

Now the synodic period, S, of the Moon (say from full Moon to full Moon) is related to the sidereal period $\mathring{P}$ by the formula I § 1.4(4),

$$\frac{1}{S} = \frac{1}{\mathring{P}} - 1.$$

Therefore, finally,

$$a - \mathring{a} = \tfrac{3}{2}\mathring{a}\,\frac{S^2}{1 + S}\,[\cos 2(M - M') - \cos 2(\delta - \delta')]. \qquad (5)$$

From this we immediately see that, in the present approximation, the secular perturbation of the semi-major axis is zero, while it has a periodic oscillation of period $\frac{1}{2}S$ and amplitude $\frac{3}{2}\mathring{a}S^2/(1 + S)$. Since a year contains about 12·3 synodic months, $S \approx 1/12{\cdot}3$ and the relative amplitude is roughly $\frac{3}{2}S^2 \approx 1\%$.

A part of this result can be interpreted in an elementary fashion. The semi-axis is at its longest at new Moon and full Moon, when the perturbations are greatest and least. Hence the period of $\frac{1}{2}S$. There is a kind of tidal influence on the orbit of the Moon. The perturbations during a complete revolution add up (approximately) to zero, or in other words, there is no secular perturbation of the semi-axis.

4. We now derive the perturbation of the vector of the areal velocity, $\boldsymbol{c}$, from the formula § 4.4(1):

$$\dot{\boldsymbol{c}} = \tfrac{1}{2}\boldsymbol{r} \times \boldsymbol{s}. \qquad (1)$$

Application of the formula 2(3) and of the trigonometrical addition theorem yields, if we put $G = 4\pi^2$,

$$\begin{aligned}\dot{\boldsymbol{c}} = 3\pi^2a^2\{\boldsymbol{i}\,.\,\tfrac{1}{2}\iota[-1 + \cos 2M + \cos 2M' - \cos 2(M - M')] +\\ + \boldsymbol{j}\,.\,\tfrac{1}{2}\iota[\sin 2M + \sin 2M' - \sin 2(M - M')] -\\ - \boldsymbol{k}\,.\,\sin 2(M - M')\}.\end{aligned}$$

Accordingly, the secular term of $\dot{\boldsymbol{c}}$ is

$$-\tfrac{3}{2}\pi^2a^2\iota\boldsymbol{i}. \qquad (2)$$

For the secular term in $\dot{\boldsymbol{c}}$ therefore, (which, for simplicity, we shall also denote by $\dot{\boldsymbol{c}}$) the formula § 4.4(4) gives:

$$\left.\begin{aligned}\dot{c} &= 0,\\ c\sin^2\iota\dot{\Omega} &= -\tfrac{3}{2}\pi^2a^2\iota\,.\,\boldsymbol{i}\,\frac{\partial\boldsymbol{k}}{\partial\Omega},\\ c\,\frac{\mathrm{d}\iota}{\mathrm{d}t} &= -\tfrac{3}{2}\pi^2a^2\iota\,.\,\boldsymbol{i}\,\frac{\partial\boldsymbol{k}}{\partial\iota}.\end{aligned}\right\} \qquad (3)$$

In 1(2) we introduced the assumption $\omega = 0$. From this and from the formulae § 4.4(2) and § 4.5(5)

$$\left.\begin{aligned}\frac{\partial \boldsymbol{k}}{\partial \Omega} &= \boldsymbol{i} \sin \iota,\\ \frac{\partial \boldsymbol{k}}{\partial \iota} &= \boldsymbol{j}.\end{aligned}\right\} \tag{4}$$

Therefore equation (3) may be written

$$\left.\begin{aligned}\dot{c} &= 0,\\ c\dot{\Omega} &= -\tfrac{3}{2}\,\pi^2 a^2,\\ \frac{\mathrm{d}\iota}{\mathrm{d}t} &= 0.\end{aligned}\right\} \tag{5}$$

Neither the magnitude of the areal velocity nor the orbital inclination, therefore, experience any secular variation (in the present approximation). The nodal line, on the other hand, moves in the opposite direction to the Moon itself, with a mean angular velocity of $\frac{3}{2}\pi^2 a^2/c$; that is, by § 4.4(5),

$$\dot{\Omega} = -\frac{3\pi^2 a^{3/2}}{\sqrt{(G\mu)}} = -\tfrac{3}{2}\,\pi\,\frac{a^{3/2}}{\sqrt{\mu}}\,. \tag{6}$$

We again eliminate μ by means of II § 3.7(7), to give

$$\dot{\Omega} = -\tfrac{3}{2}\pi P. \tag{7}$$

The period of the nodal revolution, the "Saros" of ancient Babylon determined empirically from observations of lunar and solar eclipses, is given, in the present approximation, by

$$\frac{2\pi}{|\,\dot{\Omega}\,|} = \frac{4}{3P} \tag{8}$$

or

$$\text{Saros period} = \frac{4}{3}\left(1 + \frac{1}{S}\right). \tag{9}$$

When $S \approx 1/12{\cdot}3$, this formula gives 17·7 years for the Saros period, while the observed value is 18·6 years. Our first approximation, therefore, leads to an error of about 5%. This is in agreement with our expectation that the error should be at most of the order of 8%.

In the present approximation the eccentricity also experiences no secular perturbations. This can be seen immediately from the (rigorous) formula § 4.4(5)

$$e = \left(1 - \frac{4c^2}{G\mu a}\right)^{1/2}, \tag{10}$$

since neither a nor c is subject to secular perturbations, neither is e. The periodic perturbations in e, however, are considerable. These are, indeed, the greatest of all the perturbations of the Moon, and were discovered by Hipparchus; the position of the Moon, as seen from the Earth, may be affected by as much as $1\frac{1}{4}°$.

5. Throughout the foregoing we have approximated the undisturbed Kepler motion by uniform circular motion. We cannot estimate the perturbations of the direction of perigee in this way. This can be done, however, if we represent the position vector of the Moon, $\boldsymbol{r}$, as a linear function of e in the expression 1(3) for the perturbing force $\boldsymbol{s}$. Since we can easily take into consideration the exact orbital inclination, the relative error of this approximation is of the order of $e' + e^2$, that is about $(1/60) + (1/20)^2 \approx 2\%$. A better approximation is obtained if the position vector of the Sun, $\boldsymbol{r}'$, is also represented by a linear function e'. Then the error is of the order of $e'^2 + e^2 \approx 3 \times 10^{-3}$, that is of the order of magnitude of a. In all, therefore, we must reckon with an error of about $\frac{1}{2}\%$. This holds, as well, for the perturbations in a and $\boldsymbol{c}$. The perturbations of e and $\boldsymbol{p}$, however, are mainly determined by terms which are linear in the eccentricity, so that a relative error of the order of $(e' + e) \approx 7\%$ is to be expected.

We do not give the mathematical working here, since the formulae are rather long, and no new idea is introduced. The work involved can no doubt be mastered in a few days. The use of terms of still higher orders increases the work expended so much that, in our opinion, numerical computation (especially with iteration) is to be preferred.

The chief result of this theory is that the apsidal line possesses an advancing rotation with a period of about 9·5 years.

6. The disturbing force exerted on the Sun, by the Moon, is

$$\boldsymbol{s}' = Gm\left(\frac{\boldsymbol{r} - \boldsymbol{r}'}{|\boldsymbol{r} - \boldsymbol{r}'|^3} - \frac{\boldsymbol{r}}{r^3}\right).$$

The first term in the brackets is of the order of magnitude of unity; the second is of the order of a^{-2} and is, therefore, very much larger. We can, therefore, approximate:

$$\boldsymbol{s}' = -Gm\frac{\boldsymbol{r}}{r^3}. \tag{1}$$

The appearance of this disturbing force can be understood in the following way. The Earth and the Moon move, approximately, in Kepler ellipses around their common centre of mass. This, in turn, by III § 1.4 moves in an approximate Kepler ellipse around the Sun. The perturbation of the motion of the Sun, as seen from the Earth, is nothing but the reflexion of the motion of the Earth about the centre of mass of the Earth–Moon system. It is, therefore, only a sort of perspective effect. The real gravitational perturbation (the term neglected above) is less than 10^{-5} of this apparent perturbation.

We shall now show that this explanation follows directly from the method of perturbations. If the orbits of the Sun and Moon are assumed circular, equation § 2.1(2) for the perturbation of the position vector (if the very small first term on the right-hand side is neglected) reads:

$$\begin{aligned}(\Delta\boldsymbol{r}')^{\cdot\cdot} &= \boldsymbol{s}' \\ &= -\frac{Gm}{a}\frac{\boldsymbol{r}}{a} \\ &= -\frac{Gm}{a^2}(\boldsymbol{i}\cos M + \boldsymbol{j}\sin M).\end{aligned}$$

Therefore, integrating twice with respect to t, we have

$$\Delta\boldsymbol{r}' = \frac{mP^2}{a^2}\frac{\boldsymbol{r}}{a}.$$

By Kepler's third law

$$P^2 = \mu a^3$$

(μ = Mass of the Moon, m, + mass of the Earth) it follows that

$$\Delta\boldsymbol{r}' = \frac{m}{\mu}\boldsymbol{r} = \frac{m}{\mu}a\,.\,\frac{\boldsymbol{r}}{a}. \tag{2}$$

Similarly, as in section 2, we have

$$\left.\begin{aligned} \boldsymbol{r}' &= \boldsymbol{i}' \cos M' + \boldsymbol{j}' \sin M', \\ \boldsymbol{v}' &= 2\pi(-\boldsymbol{i}' \sin M' + \boldsymbol{j}' \cos M'), \\ \frac{\boldsymbol{r}}{a} &= \boldsymbol{i}' \cos M + (\boldsymbol{j}' + \iota\boldsymbol{k}') \sin M. \end{aligned}\right\} \quad (3)$$

We split up the perturbation of the solar position vector, $\Delta\boldsymbol{r}'$, into its three components parallel to the position vector $\boldsymbol{r}'$, parallel to the velocity vector $\boldsymbol{v}'$ and parallel to the vector $\boldsymbol{k}'$ which is normal to the orbit of the Sun:

$$\Delta\boldsymbol{r}' = \alpha\boldsymbol{r}' + \Delta\lambda\frac{\boldsymbol{v}'}{2\pi} + \Delta\beta\boldsymbol{k}',$$

where $$|\boldsymbol{r}'| = \left|\frac{\boldsymbol{v}'}{2\pi}\right| = |\boldsymbol{k}'| = 1.$$

Along its orbit, therefore, the Sun experiences a change in direction of

$$\Delta\lambda = \Delta\boldsymbol{r}' \cdot \frac{\boldsymbol{v}'}{2\pi}$$

and, perpendicular to that,

$$\Delta\beta = \Delta\boldsymbol{r}' \cdot \boldsymbol{k}'.$$

From equations (2) and (3) we obtain

$$\left.\begin{aligned} \Delta\lambda &= \frac{m}{\mu} a \sin(M - M'), \\ \Delta\beta &= \frac{m}{\mu} a\iota \sin M. \end{aligned}\right\} \quad (4)$$

The empirical value of the amplitude of $\Delta\lambda$ is $6''\cdot6 = 6\cdot6/(2\cdot06 \times 10^5)$ radians. Since $a = 2\cdot57 \times 10^{-3}$, it follows that the ratio of the mass of the Moon to the total mass of the Earth and Moon together is

$$\frac{m}{\mu} = \frac{6\cdot6}{2\cdot06 \times 2\cdot57} \times 10^{-2} \approx \frac{1}{80\cdot3}.$$

By (4) the amplitude of $\Delta\beta$ is equal to the amplitude of $\Delta\lambda$ multiplied by ι, that is, multiplied by about 1/11. The observed value is, in fact, $0''\cdot6$. The period of $\Delta\lambda$ is the synodic month, and that of $\Delta\beta$ the sidereal month.

In practice, the planets, which must show similar "perspective perturbations", are observed instead of the Sun. In this way, observational errors that would arise from the thermal effect of solar radiation on the atmosphere and on the instruments are suppressed.

7. The perturbation of the perihelion of Mercury

1. The motion of Mercury shows an anomaly that cannot be explained by Newton's theory in the form in which it has so far been given. The perihelion of Mercury moves about 43″ a century faster than can be accounted for by the perturbations of all the other planets. The theory of general relativity has provided an explanation of this phenomenon. It can, however, also be explained by a small modification of Newtonian celestial mechanics, which is clearly allowable, since the law of gravity is an empirical law and has only a limited accuracy (see II § 1.3). We assume, therefore, that the expression

$$-\frac{G\mu}{r^2}\frac{\boldsymbol{r}}{r}$$

for the relative acceleration of a planet towards the Sun is only an approximation, and that the exact law has the form

$$-\frac{G\mu}{r^2}[1+\psi(r)]\frac{\boldsymbol{r}}{r}, \tag{1}$$

where $|\,\psi(r)\,| \ll 1$. The disturbing force (per unit mass) superimposed on the Newtonian gravitation is

$$\boldsymbol{s} = -\frac{G\mu}{r^2}\psi(r)\frac{\boldsymbol{r}}{r} \equiv -\sigma(r)\boldsymbol{r}. \tag{2}$$

Now, by § 4.7,

$$\dot{\boldsymbol{p}} = (\boldsymbol{v}\times\boldsymbol{r})\times\boldsymbol{s} + (\boldsymbol{s}\times\boldsymbol{r})\times\boldsymbol{v}. \tag{3}$$

Introducing (2) into (3), and remembering that

$$\tfrac{1}{2}\boldsymbol{r}\times\boldsymbol{v} = \boldsymbol{c} = c\boldsymbol{k}, \tag{4}$$

we obtain

$$\dot{\boldsymbol{p}} = 2\boldsymbol{c}\times\boldsymbol{r}\,.\,\sigma(r)$$

$$= 2c\boldsymbol{k}\times a[(\cos E - e)\boldsymbol{i} + \sin E\sqrt{(1-e^2)}\boldsymbol{j}]\,.\,\sigma[a(1-e\cos E)]$$

(see I § 6.6(5) and I § 5.6(2)) or

$$\dot{\boldsymbol{p}} = 2ca\sigma[a(1-e\cos E)][(\cos E - e)\boldsymbol{j} - \sin E\sqrt{(1-e^2)}\boldsymbol{i}]. \tag{5}$$

Now, by § 4.5(4)

$$\boldsymbol{p} = G\mu e\boldsymbol{i};$$

therefore

$$G\mu\left(\dot{e}\boldsymbol{i} + e\frac{d\boldsymbol{i}}{dt}\right) = 2ca\sigma[a(1 - e\cos E)] \times$$
$$[(\cos E - e)\boldsymbol{j} - \sin E\sqrt{(1 - e^2)}\boldsymbol{i}].$$

Taking the scalar product of this with $\boldsymbol{j}$ gives

$$e\boldsymbol{j}\cdot\frac{d\boldsymbol{i}}{dt} = \frac{2ca}{G\mu}\sigma[a(1 - e\cos E)](\cos E - e). \tag{6}$$

Now

$$\boldsymbol{j}\cdot\frac{d\boldsymbol{i}}{dt} = \dot{\omega}, \tag{7}$$

as may be immediately seen from the formula I § 5.7(6). The eccentricity e must be assumed positive, since otherwise no perihelion would be defined at all. We may, therefore, divide by e, so that it follows from equations (6) and (7) that

$$\dot{\omega} = \frac{2ca}{G\mu e}\sigma[a(1 - e\cos E)] . (\cos E - e). \tag{8}$$

It is our intention to integrate a first approximation to this equation, in which the elements on the right-hand side will be considered to be constant. To perform this integration it is convenient to introduce E as the independent variable, instead of t. We have at once

$$\dot{\omega} = \frac{d\omega}{dE}\dot{E},$$

while $\dot{E}$ is obtained by differentiating Kepler's equation,

$$E - e\sin E = n(t - T),$$

with respect to t. This gives

$$(1 - e\cos E)\dot{E} = n,$$

therefore

$$\dot{\omega} = \frac{d\omega}{dE}\frac{n}{1 - e\cos E}.$$

Introducing this expression into (8) and integrating with respect to E, we obtain for $\Delta\omega$, the change in the perihelion direction, the formula

$$\Delta\omega = \frac{2ca}{G\mu ne}\int_0^E \sigma[a(1 - e\cos E)] \,.\, (1 - e\cos E) \times (\cos E - e)\,\mathrm{d}E. \quad (9)$$

If we make use of the relations

$$2c = \sqrt{[G\mu a(1 - e^2)]}$$

and

$$n = \frac{\sqrt{(G\mu)}}{a^{3/2}},$$

equation (9) becomes

$$\Delta\omega = \frac{a^3}{G\mu e}\sqrt{(1 - e^2)}\int_0^E \sigma[a(1 - e\cos E)] \,.\, (1 - e\cos E) \times (\cos E - e)\mathrm{d}E. \quad (10)$$

Approximating

$$\sigma[a(1 - e\cos E)] = \sigma(a) - ae\cos E\sigma'(a),$$

where $\sigma'(r) \equiv \dfrac{\mathrm{d}\sigma(r)}{\mathrm{d}r}$, it follows, finally, that the secular part of $\Delta\omega$ is given by

$$(\Delta\omega)_{\text{sec}} = -\frac{1}{2}\frac{a^3}{G\mu}[3\sigma(a) + a\sigma'(a)]\,\Delta E, \quad (11)$$

where, for simplicity, we have dropped all the relatively small, terms containing e. For a complete revolution, $\Delta E = 2\pi$, so that

$$(\Delta\omega)_{\text{sec}} = -\pi\frac{a^3}{G\mu}[3\sigma(a) + a\sigma'(a)]. \quad (12)$$

Since

$$\frac{a^3}{G\mu} = \left(\frac{P}{2\pi}\right)^2$$

(II § 3.7), we can also write

$$(\Delta\omega)_{\text{sec}} = -\pi\left(\frac{P}{2\pi}\right)^2[3\sigma(a) + a\sigma'(a)]. \quad (13)$$

2. We now have to make some suitable assumption about the function $\psi(r)$, that is the factor by which we must multiply the Newtonian gravitational force to obtain the "disturbing" force. The argument of ψ must be dimensionless. One possibility is $G\mu/(\gamma^2 r)$, where γ is a constant with the dimensions of velocity. We set

$$\psi(r) = \alpha_1 \cdot \left(\frac{G\mu}{\gamma^2 r}\right) + \alpha_2 \cdot \left(\frac{G\mu}{\gamma^2 r}\right)^2 + \dots .$$

We could also have used the reciprocal $1/[G\mu/(\gamma^2 r)]$. Naturally, however, we wish to use the smallest possible number of terms, say the first only. Since the displacement of the perihelion is observed only at the smallest distances from the Sun, the approximate expression for r should be a decreasing function of r. We shall take the first term alone as sufficient, and we put

$$\psi(r) = \alpha_1 \frac{G\mu}{\gamma^2 r},$$

and

$$\sigma(r) = \alpha_1 \frac{G^2\mu^2}{\gamma^2 r^4}. \tag{1}$$

How shall we choose the constant γ? In view of the special role that the velocity of light seems to play in nature, it seems plausible to try it first as the value of γ. If it is correct, then (by analogy with most of the functions we meet in nature) we may expect $| \alpha_1 |$ to be of the order of magnitude unity, provided that $\alpha_1 \neq 0$.

Introducing equation (1) into 1(12) gives

$$(\Delta\omega)_{\text{sec}} = \alpha_1 \pi \frac{G\mu}{\gamma^2 a}. \tag{2}$$

With the data

$$\left.\begin{aligned} G &= 4\pi^2 && \text{(II § 3.8)}, \\ \mu &= 1 && \text{(II § 3.8)}, \\ \gamma &= \frac{3 \times 10^5}{4{\cdot}7} && \text{(I § 6.8, § 3.3)}, \\ a &= 0{\cdot}39 && \text{(I § 2.4)}, \end{aligned}\right\}$$

it follows that

$$\Delta\omega = \alpha_1 \times 7{\cdot}81 \times 10^{-10} \text{ radians per revolution}$$

and, therefore, for the 100/0·24 revolutions each century (see I § 1.5),

$$\alpha_1 \times 7{\cdot}81 \times 10^{-10} \times \frac{100}{0{\cdot}24} \times 206{,}264{\cdot}81 \text{ seconds of arc}$$

$$\approx \alpha_1 \times 6''{\cdot}7.$$

Therefore $\alpha_1 \approx 6$. Relativity theory gives the value

$$6\pi \frac{G\mu}{\gamma^2 a(1 - e^2)} = 42''{\cdot}9,$$

which is in very good agreement with the observed value. We have obtained only the correct order of magnitude. We cannot obtain any more than this, of course, unless we make further assumptions about the nature of the "disturbing force." The possibility remains that $\alpha_1 = 0$, so that, for example, we must put

$$\psi(r) = \alpha_2 \left(\frac{G\mu}{\gamma^2 r}\right)^2 + \dots .$$

This, however, gives the wrong order of magnitude for $(\Delta\omega)_{\text{sec}}$.

8. Lagrange's solutions of the three body problem

1. Even the analytical methods of calculating the perturbations do not provide a survey of the progress of the motion as a whole. This is possible only with very special initial conditions. The most important cases (discovered by Lagrange) will be briefly presented in this section. We start with the equations of motion of a system of three bodies given in § 2.1:

$$\ddot{\boldsymbol{r}}_1 = -G(1 + m_1) \frac{\boldsymbol{r}_1}{|\,\boldsymbol{r}_1\,|^3} - Gm_2 \left(\frac{\boldsymbol{r}_2}{|\,\boldsymbol{r}_2\,|^3} - \frac{\boldsymbol{r}_2 - \boldsymbol{r}_1}{|\,\boldsymbol{r}_2 - \boldsymbol{r}_1\,|^3} \right), \text{ etc.,}$$

and ask whether they have any solutions such that the three mass points always form an equilateral triangle.

Suppose they have such a solution, so that

$$|\,\boldsymbol{r}_1\,| \equiv |\,\boldsymbol{r}_2\,| \equiv |\,\boldsymbol{r}_2 - \boldsymbol{r}_1\,|.$$

Then the equations of motion may be written

$$\ddot{\boldsymbol{r}}_i = -G(1 + m_1 + m_2) \frac{\boldsymbol{r}_i}{|\,\boldsymbol{r}_i\,|^3}, \qquad i = 1, 2.$$

Thus, the points 1 and 2 move, relatively to the point 0, as if they were attracted by the mass $(1 + m_2)$ or $(1 + m_1)$, respectively, according to Newton's law of gravitation. So the forms of the motions, if they exist at all, are completely known, in all details. The further question, now, is whether these motions can actually occur.

Obviously this question concerns the initial conditions. Not only must the initial configuration be an equilateral triangle, but the initial velocities must also be suitably chosen. They must lie in the initial plane, and their magnitudes must be proportional to the distances of the mass points from the centre of mass. Finally, they must all make the same (oriented) angles with the corresponding position vectors. That all these conditions together are sufficient is intuitively obvious, and it can be easily proved analytically.

2. We can obtain a second Lagrangian solution if we demand that $\boldsymbol{r}_2 \equiv \lambda \boldsymbol{r}_1$, at all times, where $\lambda =$ const. $\neq 0$. If we introduce this condition into the equations of motion of the two bodies, we obtain, in general, two different expressions for $\ddot{\boldsymbol{r}}_1$. Only for particular values of λ will they be identical. Equating these expressions gives an algebraic equation for λ of the fifth degree with three real roots. The resulting equation of motion is, again, that of a system of two bodies. Conditions similar to those of the previous case are necessary for the initial configuration and velocities. Taken all together they are again sufficient.

Even for n-body systems there are solutions which are generalizations of those treated here. Thus, for instance, Maxwell took n particles of equal mass ($n \gg 1$) at the vertices of a regular n-gon as a model for the rings of Saturn. The planet itself was situated in the centre of mass. On account of their limited significance for the mechanics of the planetary system, however, we shall not discuss these solutions in any detail.

9. The *problème restreint*

1. The mass of Jupiter greatly exceeds those of the other planets, especially those of the Minor Planets and comets. Consequently the perturbations caused by Jupiter should be considered before any others. The orbit of Jupiter, however, is only slightly disturbed by the other planets. Its eccentricity is

about 1/50, so that the orbit is nearly circular. The problem of the perturbations, therefore, can be simplified by the following two idealizations:

(a) the mass of the disturbed body is negligibly small (zero);

(b) the two bodies of finite mass (the Sun and Jupiter) move around each other in circular orbits.

Similar simplifications can also be justified in the study of the motion of the satellites. In this case the disturbing body is the Sun. Since the inclination of the orbit of the Moon is small, it is sometimes possible to introduce a third assumption:

(c) the body of zero mass moves in the orbital plane of the other two bodies.

The three body problem simplified in this way is known as the *problème restreint*. Mathematically the simplification consists in the reduction of the original system of 18 first order differential equations to one of only 6 or even 4. By § 1.1 these equations can be written in absolute coordinates:

$$\ddot{\boldsymbol{r}} = Gm_1 \frac{\boldsymbol{r}_1 - \boldsymbol{r}}{|\,\boldsymbol{r}_1 - \boldsymbol{r}\,|^3} + Gm_2 \frac{\boldsymbol{r}_2 - \boldsymbol{r}}{|\,\boldsymbol{r}_2 - \boldsymbol{r}\,|^3}\,.$$

The indices 1 and 2 denote the two bodies of finite mass, while $\boldsymbol{r}$ is the position vector of the disturbed body. If the origin of the coordinates is taken as the centre of mass, then, by § 1.3,

$$m_1\boldsymbol{r}_1 + m_2\boldsymbol{r}_2 = 0.$$

The vectors $\boldsymbol{r}_1$ and $\boldsymbol{r}_2$ are known functions of the time. If the orbital plane is taken as the (x, y)-plane, then

$$\left.\begin{aligned} x_i &= \mathring{x}_i \cos nt - \mathring{y}_i \sin nt, \\ y_i &= \mathring{x}_i \sin nt - \mathring{y}_i \cos nt, \\ z_i &= 0, \end{aligned}\right\} \qquad i = 1, 2.$$

The positions of the two mass points at the time $t = 0$ are given by $\mathring{\boldsymbol{r}}_1$ and $\mathring{\boldsymbol{r}}_2$; their mean motion, around each other, is denoted by n.

Now we can let the coordinate system rotate about the z-axis with the angular velocity n, so that it rotates with the two finite masses. We then have a new x'-axis,—the line which joins these two bodies, and $y_i' \equiv 0$. We introduce, therefore, a new (x', y', z')-coordinate system, which is defined by the equations

$$\left.\begin{aligned} x &= x' \cos nt - y' \sin nt, \\ y &= x' \sin nt + y' \cos nt, \\ z &= z' \end{aligned}\right\}.$$

After having made the transformation we can drop the primes. The equations of motion then become

$$\ddot{\boldsymbol{r}} + 2n\begin{pmatrix} -\dot{y} \\ \dot{x} \\ 0 \end{pmatrix} = n^2\begin{pmatrix} x \\ y \\ 0 \end{pmatrix} + Gm_1 \frac{\boldsymbol{r}_1 - \boldsymbol{r}}{|\boldsymbol{r}_1 - \boldsymbol{r}|^3} + Gm_2 \frac{\boldsymbol{r}_2 - \boldsymbol{r}}{|\boldsymbol{r}_2 - \boldsymbol{r}|^3}. \quad (1)$$

The vectors $\boldsymbol{r}_1$ and $\boldsymbol{r}_2$ are now constants. The time variable t no longer appears explicitly in the equations.

We now introduce dimensionless variables into these equations of motion. We write

$$\left.\begin{aligned} m_i' &= \frac{m_i}{m_1 + m_2}, \qquad i = 1, 2, \\ \boldsymbol{r}' &= \boldsymbol{r}/a, \\ \boldsymbol{r}_i' &= \boldsymbol{r}_i/a, \qquad i = 1, 2, \\ t' &= \left(G\frac{m_1 + m_2}{a^3}\right)^{1/2} t, \end{aligned}\right\}$$

where

$$a = |\boldsymbol{r}_2 - \boldsymbol{r}_1|;$$

we recall also the relation II § 3.7(7)

$$n^2 = G\frac{m_1 + m_2}{a^3}.$$

Then the equations of motion read, in the dimensionless variables,

$$\ddot{\boldsymbol{r}} + 2\begin{pmatrix} -\dot{y} \\ \dot{x} \\ 0 \end{pmatrix} = \begin{pmatrix} x \\ y \\ 0 \end{pmatrix} + m_1 \frac{\boldsymbol{r}_1 - \boldsymbol{r}}{|\boldsymbol{r}_1 - \boldsymbol{r}|^3} + m_2 \frac{\boldsymbol{r}_2 - \boldsymbol{r}}{|\boldsymbol{r}_2 - \boldsymbol{r}|^3},$$

where, for simplicity in writing, we have again dropped the primes.

2. These equations of motion possess a first integral, named after Jacobi, which corresponds to the energy integrals of II § 3.3

and III § 1.6. It is also obtained in a quite analogous manner, namely, by taking the scalar product of the equation of motion 1(1) with $\dot{\boldsymbol{r}}$, and then integrating with respect to t. This gives

$$\tfrac{1}{2}\boldsymbol{v}^2 - \mathscr{U}(\boldsymbol{r}) = -g, \tag{1}$$

where g denotes a constant, and

$$\mathscr{U}(\boldsymbol{r}) \equiv \tfrac{1}{2}(x^2 + y^2) + \frac{m_1}{|\boldsymbol{r} - \boldsymbol{r}_1|} + \frac{m_2}{|\boldsymbol{r} - \boldsymbol{r}_2|}. \tag{2}$$

The value of the constant g is specified by the initial conditions.

Since $\boldsymbol{v}^2 \geqslant 0$, the orbital curve must obviously belong to the closed region, $\mathscr{C}_g$, which is defined by the inequality

$$\mathscr{U}(\boldsymbol{r}) \geqslant g.$$

If g is sufficiently large, then $\mathscr{C}_g$ possesses at least one bounded component, as well as an unbounded one. That is to say $\mathscr{C}_g$ is the union of several distinct closed subsets (the so-called components), of which at least one is bounded and one is unbounded. To show this, let K_δ be the surface of a sphere of radius $\delta > 0$, say around the point $\boldsymbol{r}_1$. The continuous function $\mathscr{U}(\boldsymbol{r})$ has a maximum A_δ on K_δ. We now choose $g > A_\delta$ so that, since $\mathscr{U}(\boldsymbol{r}) \to \infty$ as well for $\boldsymbol{r} \to \boldsymbol{r}_1$ as for $(x^2 + y^2) \to \infty$, there are certainly points $\boldsymbol{r}$, both inside and outside K_δ, for which $\mathscr{U}(\boldsymbol{r}) \geqslant g$. On K_δ itself, however, $\mathscr{U}(\boldsymbol{r}) < g$, so that in the neighbourhood of K_δ there are no points belonging to $\mathscr{C}_g$. The sphere K_δ, therefore, surrounds one component of $\mathscr{C}_g$. If, now, the starting point of the motion is within such a component, then the whole orbit must lie within it. The particle of "infinitesimal" mass must, therefore, always stay in the neighbourhood of the particular mass point.

This condition is fulfilled, for example, for the Moon. It cannot be concluded, however, that the Moon must always remain near the Earth. This particular conclusion would be true only if the mass of the Moon were exactly zero. It ceases to be valid for any positive mass, no matter how small. It follows from the theory of differential equations, however, that, in a bounded time interval, the Moon remains arbitrarily near to the fictitious body of infinitesimal mass—and, of course, the smaller the mass of the Moon, the longer it will do so. The length of this interval of time does not seem to have been estimated.

The exact calculation of the quantity

$$A_\delta = \operatorname*{Max}_{|r-r_1|=\delta} \mathscr{U}(\boldsymbol{r})$$

is troublesome. We substitute for it a simple estimate. Obviously the quantity

$$A'_\delta = \operatorname*{Max}_{|r-r_1|=\delta} [\tfrac{1}{2}(x^2+y^2)] + \frac{m_1}{\delta} + \\ + \operatorname*{Max}_{|r-r_1|=\delta}\left[\frac{m_2}{|\boldsymbol{r}-\boldsymbol{r}_2|}\right]$$

is greater than A_δ. Let us, therefore, choose $g \geqslant A'_\delta$, so that g is certainly greater than A_δ, and the existence of a bound component of $\mathscr{C}_g$, containing $\boldsymbol{r}_1$, is assured. We determine A'_δ as follows. The mass points 1 and 2 lie on the x-axis, so that their abcissae are $-m_2$ and m_1. Therefore

$$\left.\begin{aligned} \operatorname*{Max}_{|r-r_1|=\delta} [\tfrac{1}{2}(x^2+y^2)] &= \tfrac{1}{2}(m_2+\delta)^2, \\ \operatorname*{Max}_{|r-r_1|=\delta}\left[\frac{m_2}{|\boldsymbol{r}-\boldsymbol{r}_2|}\right] &= \frac{m_2}{1-\delta} \end{aligned}\right\}$$

and

$$A'_\delta = \tfrac{1}{2}(m_2+\delta)^2 + \frac{m_1}{\delta} + \frac{m_2}{1-\delta}.$$

In particular, for $\delta = m_1$ we have $A'_\delta = \frac{5}{2}$. Each set $\mathscr{C}_g$, with $g \geqslant \frac{5}{2}$, therefore, possesses one component inside each of the spheres centred on the two mass points and of radii m_1 and m_2 respectively. For very small m_1, δ and m_1/δ, (the Moon!) we have $A_\delta \approx A'_\delta \approx \frac{3}{2}$.

3. The Lagrangian solutions are, of course, also found in the *problème restreint*. In the rotating system of coordinates they are represented as five fixed points—the "points of libration". In the planetary system one of these solutions is approximately realized. The Trojan group of Minor Planets always form a roughly equilateral triangle with Jupiter and the Sun. The stability of this configuration has been investigated by means of the method of small oscillations of § 3.8. It proves to be stable, if the ratio of the two finite masses is very small (or very large).

As we know from § 3.8, however, no final conclusion about stability can be obtained by this method. It gives an approximate representation, however, over a finite interval of time, for those motions whose initial conditions are nearly Lagrangian. The fact that the solutions of the variational equations are of a stable type shows that both motions stay together over such intervals.

The collinear solution also appears to be realized in nature in the case of the Sun, the Earth and the "Gegenschein". This is supposed to be an accumulation of meteors in the opposite direction to the Sun. The method of small oscillations, however, shows this solution to be unstable.

The simplification of the *problème restreint* has its chief application in the theory of the motion of the Moon. The general principles of the perturbation theory are the same as for the planets. There are many purely mathematical investigations into the existence of periodic motions, which we shall not discuss any further.

10. Summary

Ten first integrals are known for the equations of motion of an n-body system in absolute space. They are the three integrals of momentum, three of the centre of gravity, three of the moment of momentum, and the energy integral. The transition to relative coordinates lowers the order of the system of equations by six, since the momentum and centre of gravity integrals are thereby eliminated (§ 1).

If the Kepler motion is taken as the zero approximation to the motions of the planets, then the iterative method yields, even after the first step, an approximation which is often sufficiently good to compare with observation. (This is especially true of the Major Planets.) A Fourier expansion is suitable for the computation. It makes possible the distinction between secular, long and short periodic perturbations. The "perturbations in the elements" can be computed from the "perturbations in the coordinates" (§ 2).

The perturbations in the elements can also be evaluated directly, without a detour through the perturbations in the coordinates. For this the method of the "variation of the parameters" of the differential equations is used. The parameters may be varied either in solution functions of the undisturbed system, (Lagrange's method) or in their first integrals (Poisson's method). If the system of differential equations contains a small parameter (and possesses the required properties of regularity), then the solution can be developed in powers of the parameter. The method can also be adapted to the case in which the initial conditions are disturbed instead of the right-hand sides of the differential equations (the "force"). By means of the expansion in powers of the small parameter, the original system of equations can be solved in an infinite sequence of inhomogeneous linear systems. For the perturbations of the initial conditions, however, the first equation of this sequence is homogeneous. All inhomogeneous equations have the same homogeneous part, the "Jacobi variational equation". For questions of stability this has constant coefficients, and then possesses trigonometrical or exponential solutions. The solution of the inhomogeneous system is obtained from that of the homogeneous system by the variation of parameters. A generalization of the system with constant coefficients is that with periodic coefficients. Finally, it was shown by an example, that the "Method of Small Oscillations" can lead to false results in questions of stability (§ 3).

Poisson's method is especially appropriate to the formation of the perturbation equations of celestial mechanics. The equations can be solved by successive approximations or by expanding in powers of the small masses (§ 4).

The explicit presentation of the solutions of the perturbation equations would be very troublesome. As for the perturbations of the coordinates, so perhaps here also a partly analytical and partly numerical procedure is the most suitable. Simple explicit approximations display the qualitative character of the solutions (over a bounded interval of time) and allow the orders of magnitude to be estimated, while numerical computation of exact predictions of the motion makes comparison between theory and observation possible. In particular, simple (though rather rough) analytical first approximations can be given for the satellites, on account of their very small distances from their primaries (§§ 5 and 6).

Application of the theory of perturbations shows that the motion of the perihelion of Mercury can be explained by a slight modification of Newton's law of gravitation at small distances. The correct order of magnitude is at once obtained if it is assumed that the velocity of light is in some way associated with gravity. (How it may be associated is not specified any more closely). (§ 7.)

Even the theory of perturbations cannot give a general survey of the motion as a whole. This is only possible for special initial conditions, e.g. the Lagrangian solutions (§ 8).

Finally, the *problème restreint* considers a three body system in which the mass of one of the bodies is supposed to be vanishingly small. Thus its motion is determined by the other two bodies, but it exerts no reaction on them. The relative motion of the two massive bodies is assumed to be circular. This simplified formulation of the equations of motion is particularly useful in the theory of the motion of the Moon (§ 9).

Exercises

1. Apply the generalized centre of gravity law of § 1.4 to the Earth–Moon system in its motion around the Sun. Assuming uniform circular and coplanar motion, discuss, quantitatively and qualitatively, the influence of the motion of the Moon on the apparent annual motion of the Sun (see § 6.6).

2. Derive the law of areas of the two body system (II § 3.1) from the generalized law of the conservation of moment of momentum (III § 1.6).

3. Derive the absolute coordinates of the planets (referred to the centre of mass) from their relative coordinates (referred to the Sun) (§ 1.9).

4. Prove that the total moment of momentum of the planetary system, $\sum_0^n \boldsymbol{r}_i \times m_i \boldsymbol{v}_i$ is equal to $\sum_1^n \mathbf{r}_i' \times m_i \boldsymbol{v}_i'$, up to terms in the second order of the planetary masses $m_i \ll 1$, where $\boldsymbol{r}_i' = \boldsymbol{r}_i - \boldsymbol{r}_0$ denotes the position vectors relative to the Sun.

5. Apply the last theorem to the case of two planets in the following way. Take the "invariable plane", that is the plane through the centre of mass perpendicular to the total moment of momentum, as the (x, y)-plane of the coordinate system. (The invariable plane may here be taken through the Sun.) Then the first two components of the vectors $\sum_1^2 \boldsymbol{r}_i' \times m_i \boldsymbol{v}_i'$ are equal to zero, to terms of the second order in m_1 and m_2. These components may be expressed in terms of the instantaneous osculating elements (II § 3.2, III § 2.2 and 4). Prove that if the two orbital inclinations differ from zero, then the longitudes of their nodes differ by approximately 180°. (Jacobi's law of nodes. For relative coordinates it is only approximately true; for other, suitably defined coordinates it holds exactly.)

6. Prove Jacobi's law of the nodes by taking the vector product of the equation

$$\boldsymbol{r}'_1 \times m_1 \boldsymbol{v}'_1 + \boldsymbol{r}'^{1}_{2} \times m_2 \boldsymbol{v}'_2 = \textbf{const.}$$

with the constant vector of its right-hand side.

7. In the last equation, express the third component in terms of the elements and prove the following. If at time $t = 0$ both planets are revolving in the same direction, and their orbits have very small inclinations and eccentricities, then the inclinations and eccentricities will remain small, as long as the semi-major axis is not (appreciably) altered. (Laplace's stability theorem.)

8. Prove that for two coplanar circular orbits the secular term in the semi-major axis vanishes, in the first approximation (§ 4.1). (According to the stability theorem of Lagrange and Laplace this also holds under more general assumptions.)

9. Explain, qualitatively, the effects of the perturbations of two coplanar circular orbits in Cartesian coordinates (§ 2.1).

10. Assume the undisturbed orbits of Uranus and Neptune to be coplanar, and explain in detail how the observed perturbations $\Delta\lambda_1$ of Uranus depend on the semi-major axis a_2, the mass m_2 and the phase $\mathring{\theta}_2$ of Neptune (§ 2.5).

11. Integrate the equation of motion

$$\dot{x} = -x + \alpha x^2, \quad |\alpha| \ll 1,$$

a. by means of Lagrange's method of variation of parameters,
b. by means of Poisson's method of variation of parameters,
c. by expansion in powers of the parameter α (§ 3.1–3).

12. Apply the matrix method of § 3.5 to the differential equation $\dot{\boldsymbol{x}} = \boldsymbol{x}$.

13. Solve the system of differential equations

$$\dot{\boldsymbol{x}} = \boldsymbol{A}\boldsymbol{x} + \boldsymbol{b}(t), \qquad \boldsymbol{A} = \textbf{const.},$$

by means of Poisson's variation of the parameters (§ 3.9).

14. What is the order of magnitude (absolute and relative) of the right-hand side of the perturbation equations of § 4, if the disturbed body is the Moon, and the undisturbed body the Sun? Estimate the corresponding orders of magnitude in the mutual perturbations of Jupiter and Saturn. The comparison of these two orders of magnitude shows the particular difficulties of lunar theory, especially the perturbations in the observed direction.

15. In equations § 4.7(5) and (11), the required gradients are multiplied by factors of $e \sin E$ and e respectively. Do these lead to singularities for the gradients when $e = 0$ or $E = 0$?

16. Complete the formation of the differential equation for T (§ 4.7).

17. As is known from equation § 4.7(2), terms of the form $t \sin nt$ appear in the differential equation for T. Since they are troublesome (oscillations with growing amplitude!), they are usually removed by a transformation of

variables. Discuss the significance of this procedure in view of § 2.3 and 2.5.

18. Why does the periodic repetition of solar and lunar eclipses (the Saros) coincide with the period of the Moon's nodes?

19. The perturbations of the Sun, by the Moon (§ 6.6), $\Delta\lambda$, $\Delta\beta$ are determined empirically not through solar, but through planetary observations. Derive the formulae from which the solar perturbations can be obtained from the corresponding planetary perturbations. Which planets are most suitable? (Remember to consider the mutual perturbations of the planets!)

20. How large is the motion of the perigee of the Moon (§ 7.2(2))?

21. Generalize the Lagrange triangular solution of § 8.1 to the case of a regular n-gon. Each corner is occupied by a point of mass m. The centre may contain an arbitrary mass.

22. Set up the equation of the fifth degree mentioned in § 8.2 and discuss the number and position of the real roots.

23. Estimate the numerical value of the energy constant g in § 9.2 for the Moon. Show that its orbit would lie within a closed surface surrounding the Earth if the mass of the Moon were zero.

24. Show that the triangular solution of the *problème restreint* is "stable" for a sufficiently large ratio of the two finite masses, m_1 and m_2 (§ 9.3).

NOTES

§ **1.1.** see Kurth [2].

§ **1.2.** On collisions see Wintner [7] or

[19] Siegel, C. L., *Vorlesungen über Himmelsmechanik*, Berlin–Göttingen–Heidelberg, 1956.

§ **2.3.** On almost periodic functions see

[20] Bohr, H., *Fastperiodische Funktionen*, (Ergebnisse der Mathematik, Springer, 1932.)

§ **2.5.** For the history of the discovery of Neptune, see the very fascinating account by

[21] Smart, W. M., Supplements to Monthly Notices, 1946.

§ **3.2.** Note that Poisson's method yields the differential equations in a form suitable for integration (namely, solved for the time derivative), but it assumes more than Lagrange's method. The latter requires only the general solution of the undisturbed problem. Poisson's method, however, also needs the corresponding first integrals of the undisturbed problem.

§ **3.7.** For the case of equal characteristic roots see Wintner [7]. The theory of elementary divisors needed here will be found in a very readable form in

[22] Bôcher, M., *Introduction to Higher Algebra*, New York, 1927.

§ **4.1.** Many authors ascribe a fundamental significance to Hamilton–Jacobi theory, in celestial mechanics. I cannot agree with them. Of course the theory presents the mechanism of very general coordinate transformations clearly and beautifully; it appears to me, however, to be of limited (or even no) use for the solution of concrete problems. The same results can be obtained quite directly and much more simply by Poisson's method (which has hitherto been neglected). The reputation of the Hamilton–Jacobi theory appears to me to have an historical rather than an objective foundation. That it should be tried is understandable: we seek first integrals and these can easily be found if the coordinates are suitably chosen. (An example is the use of polar coordinates in problems with "spherical symmetry".) That the theory was retained, in spite of failures, is a phenomenon which, in my opinion, is to be explained psychologically.

§ **4.2 *et seq.*** Some authors expand the "disturbing function" (the potential function of the disturbing force) in some series before they set up the differential equations. They do not distinguish quite so sharply, therefore, between the setting up of the equations and their solution. It seems more suitable to me first to put the equations in the most explicit form possible, and only then to study the possibility of their practical solution.

The disturbing function just mentioned plays no rôle in our work. It entered only because the representation of the force as the (negative) gradient of the potential was necessary for the application of Hamilton–Jacobi theory. It is not necessary, however for Poisson's method.

With the help of this function, the differential equations for the disturbed elements are often written in a half explicit form, in which the right-hand sides still contain the partial derivatives with respect to the disturbed elements. Though this seems formally fairly simple, a whole series of further analytical transformations is needed before the right-hand sides can be integrated.

§ **4.6.** I have no practical experience in the integration of the perturbation equations. My proposals for the computation of the solution, therefore, indicate only what I suppose should be tried.

§ **6.3 *et seq.*** On lunar theory see Plummer [6] or

[23] Brown, E. W., *An Introductory Treatise on Lunar Theory*, Cambridge, 1896.

A simple qualitative treatment is given by Moulton [24].

[24] Moulton, F. R., *An Introduction to Celestial Mechanics*. New York, 1947.

§ **7.2.** The relativistic solution can be found in, e.g.,

Eddington [14], Bergmann [15], Laue [16].

§ **8.** The Lagrangian solutions can be generalized—see Wintner [7]. Corresponding solutions also hold for arbitrary forces, see

[25] Kurth, R., On Lagrange's Triangular Solution of the Problem of Three Bodies, *Archiv. d. Math.*, **8**, 282, 1957.

§ **9.2.** More detailed discussions of "Hill's Surfaces" are given, e.g., by Wintner [7] or Plummer [6]. Both also give a detailed account of Lunar Theory on the basis of the *problème restreint.*

CHAPTER IV

THE PLANETS AND THE MOON AS RIGID BODIES

1. Perturbations of the orbits of satellites caused by flattening of the planets

1. Hitherto we have considered all the heavenly bodies as mass points. In this chapter, however, we shall consider their extension in space. This can be done in two ways, either the body is regarded as a system of a finite number of mass points, or as a material continuum. In this book we shall deal only with the first kind of representation, since the methods developed here may be readily applied to it, while the investigation of gravitational continua requires more difficult mathematical methods, as well as a more detailed consideration of the physical state of the material.

First of all we shall investigate the force which a planet constructed from a finite number of mass points would exert on a distant mass point (a satellite). Let m be the mass of the satellite, and $\boldsymbol{r} = r\boldsymbol{u}$ be its position vector referred to the mass centre of the planet, ($r = |\,\boldsymbol{r}\,|$). The position vectors of the n particles of masses m_i ($i = 1, 2, \ldots, n$) which together form the planet are $\boldsymbol{r}_i = r_i\boldsymbol{u}_i$, where $r_i = |\,\boldsymbol{r}_i\,|$. Since we take the mass centre of the planet as the origin of coordinates, we have

$$\sum_{i=1}^{n} m_i\boldsymbol{r}_i = 0. \tag{1}$$

Then, by III § 1.1 the force $\boldsymbol{f}$ exerted on the satellite by the planet is

$$\begin{aligned}\boldsymbol{f} &= Gm \sum_{i=1}^{n} m_i \frac{\boldsymbol{r}_i - \boldsymbol{r}}{|\,\boldsymbol{r}_i - \boldsymbol{r}\,|^3} \\ &= Gm \sum_i m_i(\boldsymbol{r}_i - \boldsymbol{r})(\boldsymbol{r}_i^2 - 2r_ir\boldsymbol{u}_i\,.\,\boldsymbol{u} + \boldsymbol{r}^2)^{-3/2}\end{aligned} \tag{2}$$

from which follows, by the binomial theorem,

$$\begin{aligned}\boldsymbol{f} = Gm \sum_i m_i(\boldsymbol{r}_i - \boldsymbol{r})\Big[1 &+ \frac{3r_i}{r}\boldsymbol{u}_i\,.\,\boldsymbol{u} - \\ &- \frac{3}{2}\left(\frac{r_i}{r}\right)^2 + \frac{15}{2}\left(\frac{r_i}{r}\right)^2 (\boldsymbol{u}_i\,.\,\boldsymbol{u})^2\Big],\end{aligned} \tag{3}$$

where we have neglected terms of the third and higher orders in $(r_i/r) \ll 1$. From (1) and (3) it follows that

$$f = -\frac{Gm \sum_i m_i}{r^2} u + \frac{Gm}{r^4}\left\{\frac{3}{2} u \sum_i m_i r_i^2[1 - 5(uu_i)^2] + \right.$$
$$\left. + 3 \sum_i m_i r_i^2 (uu_i) u_i \right\}. \quad (4)$$

The disturbing force s per unit mass, on the satellite, is, therefore,

$$s = \frac{3}{2}\frac{G}{r^4}\left[u \sum_i m_i r_i^2 + 2 \sum_i m_i r_i^2 (u \, . \, u_i) u_i - \right.$$
$$\left. - 5u \sum_i m_i r_i^2 (uu_i)^2 \right]. \quad (5)$$

To express this more clearly, we denote the components of r_i by x_i, y_i, z_i, and write

$$J = \sum_i m_i \begin{pmatrix} x_i x_i & x_i y_i & x_i z_i \\ y_i x_i & y_i y_i & y_i z_i \\ z_i x_i & z_i y_i & z_i z_i \end{pmatrix} \quad (6)$$

and

$$3J = \sum_i m_i(x_i^2 + y_i^2 + z_i^2) = \sum_i m_i r_i^2. \quad (7)$$

Then, as the reader may convince himself by working it out, equation (5) may be written in the form:

$$s = \frac{1}{2}\frac{G}{r^4}\{3Ju + 2J - 5(u'Ju)U\}u, \quad (8)$$

where U is the unit matrix, and u' is the transpose of u (that is the vector u written as a row).

The quadratic form in u

$$u'Ju = \sum_i m_i(r_i u)^2 \quad (9)$$

is, by definition, positive definite and may therefore be geometrically represented by an ellipsoid. If we choose the principal axes of this ellipsoid as coordinate axes, then

$$J = \begin{pmatrix} J_x & 0 & 0 \\ 0 & J_y & 0 \\ 0 & 0 & J_z \end{pmatrix}, \quad (10)$$

where

$$J_x = \sum_i m_i x_i^2. \qquad \text{etc.} \quad (11)$$

If the planet has spherical symmetry, then

$$J_x = J_y = J_z = J \tag{12}$$

and, therefore,

$$\boldsymbol{J} = J\boldsymbol{U}, \tag{13}$$

so that in this case, by (8), $\boldsymbol{s} = 0$. The disturbing force, therefore, vanishes if the planet is spherically symmetrical: the planet then attracts the satellite as if all its mass were concentrated in its mass centre. We have shown here only that the terms of the fourth order in r vanish; that all the terms of higher order also vanish can be more simply deduced from a study of the continuous model—which we are not considering.

In fact, the planets have only approximately the form of a sphere (and supposedly, therefore, a corresponding internal structure). Consequently we introduce the matrix

$$\boldsymbol{\Delta} = \boldsymbol{J} - J\boldsymbol{U}, \tag{14}$$

and assume that all the elements of $\boldsymbol{\Delta}$ are very much smaller in magnitude than J. By means of (14) we can transform the expression (8), for the disturbing force, into the following, simpler form:

$$\boldsymbol{s} = \frac{G}{r^4}\{\boldsymbol{\Delta u} - \tfrac{5}{2}(\boldsymbol{u}'\boldsymbol{\Delta u})\boldsymbol{u}\}. \tag{15}$$

2. If this last expression is introduced into the perturbation equations of III § 4, then the corresponding perturbations can be calculated—assuming that the matrix $\boldsymbol{\Delta}$ is known. In practice $\boldsymbol{\Delta}$ is never known *a priori* but must, just in reverse, be evaluated from the observed perturbations of the satellites. It is more convenient to write the matrix with its elements referred to a coordinate system whose axes are the principal axes of the ellipsoid representing the quadratic form $\boldsymbol{u}'\boldsymbol{\Delta u}$, rather than to our usual coordinate system (in which the (x, y)-plane is a mean plane of the orbit of the Earth, see I § 4.1).

At this point we must make some hypothesis about the physical properties of the planets. In general $\boldsymbol{\Delta}$ will vary with the time, since the physical state of the planet may also depend on time. It is simplest to regard the planets as rigid bodies, that is the mutual distances of its particles are assumed to be the same at all times. For the Earth, for example, this is undoubtedly a good approximation. It implies that all the elements of the diagonal form equation 1(10) of the matrix $\boldsymbol{J}$, (and also, therefore, of the diagonal form of the matrix $\boldsymbol{\Delta}$) are constant with respect to time. The directions of the axes of the ellipsoid may, however, vary with the time. Now the planets rotate (see § 2), and they possess an axis of symmetry which (nearly) coincides with the axis of rotation, which has a (nearly) fixed direction in space (i.e. relative to the fixed stars). We can, therefore, introduce a coordinate system fixed in space, in which $\boldsymbol{J}$ has a diagonal form: as z-axis we take the axis of rotation, the x- and y-axes then lie in the equatorial plane of the planet; we choose them to be fixed in space. Then $\boldsymbol{J}$ and $\boldsymbol{\Delta}$ are still diagonal matrices, since the expression $\boldsymbol{u}'\boldsymbol{\Delta u}$ has the form

$$[\Delta_x(u_x^2 + u_y^2) + \Delta_z u_z^2]$$

in the coordinate system fixed relatively to the body, and therefore also has this form in the system fixed in space just described.

3. To give a simple application of this theory, we calculate the secular rotation of the apsidal line for a satellite which moves in the equatorial plane of a rigid planet possessing rotational symmetry. Let us choose this plane as the (x, y)-plane, then

$$\boldsymbol{u} = \begin{pmatrix} u_x \\ u_y \\ 0 \end{pmatrix} \tag{1}$$

and

$$\begin{aligned} \boldsymbol{\Delta u} &= \begin{pmatrix} \Delta_x & 0 & 0 \\ 0 & \Delta_y & 0 \\ 0 & 0 & \Delta_z \end{pmatrix} \begin{pmatrix} u_x \\ u_y \\ u_z \end{pmatrix} = \begin{pmatrix} \Delta_x u_x \\ \Delta_y u_y \\ 0 \end{pmatrix} \\ &= \Delta_x \boldsymbol{u} \end{aligned} \tag{2}$$

(since $\Delta_x = \Delta_y$). To express this in the usual notation, we introduce the so-called principal moments of inertia A, B, C, defined by

$$A = \sum_i m_i(y_i^2 + z_i^2) = J_y + J_z, \text{ etc.} \tag{3}$$

Hence,

$$\begin{aligned} \Delta_x &= J_x - J && \text{(by 1(14))} \\ &= J_x - \tfrac{1}{3}(J_x + J_y + J_z) && \text{(by 1(7))} \\ &= \tfrac{1}{3}[(J_x - J_y) + (J_x - J_z)] \\ &= \tfrac{1}{3}[(B - A) + (C - A)]; && \text{(by (3)),} \end{aligned} \tag{4}$$

for $\Delta_x = \Delta_y$, that is for $A = B$:

$$\Delta_x = \tfrac{1}{3}(C - A).$$

Equation (2), therefore, reads

$$\boldsymbol{\Delta u} = \tfrac{1}{3}(C - A)\boldsymbol{u}, \tag{5}$$

and by 1(15) the disturbing force, in this case, reduces to

$$\boldsymbol{s} = -\sigma(r) \, . \, \boldsymbol{r}, \tag{6}$$

where

$$\sigma(r) = \tfrac{1}{2} \frac{G}{r^5} (C - A). \tag{7}$$

By III § 7.1(5) the secular displacement of periastron, per period, is given by

$$\Delta\omega = -\frac{\pi a^3}{G\mu}\left[3\sigma(a) + \frac{a \, \mathrm{d}\sigma(a)}{\mathrm{d}a}\right]; \tag{8}$$

so, in this case,

$$\Delta\omega = \pi \frac{C - A}{\mu a^2} \tag{9}$$

per revolution. By means of this formula the value of $(C - A)$ may be found from the empirical values of $\Delta\omega$, μ, a, and from this value some conclusions may be drawn about the internal structure of the planet.

4. There is another way of obtaining information about the quantity $(C - A)$, namely through observations of the directions of the nodes of the satellite orbit. We take the equatorial plane of the planet as the (x, y)-plane, and assume that the inclination, ι, of the orbit of the satellite is so small that ι^2 may be neglected in comparison with unity. We further limit ourselves to the consideration of circular orbits. Then, by III § 6.2(5),

$$\boldsymbol{u} = \begin{pmatrix} \cos M \\ \sin M \\ \iota \sin M \end{pmatrix}. \tag{1}$$

Thus, for the disturbing force

$$\boldsymbol{s} = \frac{G}{r^4}[\boldsymbol{\Delta u} - \tfrac{5}{2}(\boldsymbol{u}'\boldsymbol{\Delta u})\boldsymbol{u}], \tag{2}$$

we have

$$\boldsymbol{s} = \frac{G}{a^4}[-\tfrac{3}{2}\Delta_x \boldsymbol{u} + \iota(\Delta_z - \Delta_x) \sin M \, . \, \boldsymbol{k}]. \tag{3}$$

Therefore, from III § 4.4(1)

$$\dot{\boldsymbol{c}} = \tfrac{1}{2}\boldsymbol{r} \times \boldsymbol{s},$$

it follows that

$$\dot{\boldsymbol{c}} = \frac{1}{2}\frac{G\iota}{a^3}(\Delta_z - \Delta_x)\begin{pmatrix} \sin^2 M \\ -\cos M \sin M \\ 0 \end{pmatrix}, \tag{4}$$

while by III § 4.4(3):

$$c\left(\frac{\partial \boldsymbol{k}}{\partial \Omega}\right)^2 \dot{\Omega} = \dot{\boldsymbol{c}}\,\frac{\partial \boldsymbol{k}}{\partial \Omega}. \tag{5}$$

Now, by III § 6.4(4), and under the above assumptions,

$$\frac{\partial \boldsymbol{k}}{\partial \Omega} = \boldsymbol{i}\iota, \tag{6}$$

and

$$c = \tfrac{1}{2}\sqrt{(G\mu a)}. \tag{7}$$

Introducing (4), (6) and (7) into (5) we obtain

$$\sqrt{(G\mu a)}\,\dot{\Omega} = \frac{G}{a^3}(\Delta_z - \Delta_x)\sin^2 M$$

and, for the secular term of $\dot{\Omega}$:

$$\sqrt{(G\mu a)}\,\dot{\Omega}_{\text{sec}} = \frac{1}{2}\frac{G}{a^3}(\Delta_z - \Delta_x). \tag{8}$$

The rotation of the nodal line during a revolution, therefore, is given by

$$\sqrt{(G\mu a)}\;\Delta\Omega = \frac{1}{2}\frac{G}{a^3}(\Delta_z - \Delta_x)P. \tag{9}$$

Now, remembering that,

$$P = 2\pi\frac{a^{3/2}}{\sqrt{(G\mu)}}, \tag{10}$$

(by II § 3.4(7)) and

$$\begin{aligned} \Delta_z - \Delta_x &= (J_z - J) - (J_x - J) \\ &= (J_z - J_x) \\ &= (J_z + J_y) - (J_x + J_y) \\ &= A - C \end{aligned} \tag{11}$$

(by 3), we can transform (9) into our final formula:

$$\Delta\Omega = -\pi\frac{C - A}{\mu a^2} \tag{12}$$

per revolution. Thus $\Delta\Omega$ is just the opposite value of the secular perturbation, $\Delta\omega$, of the apsidal line.

2. The rotation of the Earth

1. Not only are the motions of satellites disturbed by the flattening of their planets, but the satellites themselves disturb the rotation of their planets through this same flattening. Consider a planet constructed of n particles, of masses m_1, m_2, ..., m_n, with position vectors $\boldsymbol{r}_1 = r_1\boldsymbol{u}_1$ etc. The force exerted on the ith particle by the Moon (or the Sun) is denoted by $\boldsymbol{g}_i$. Then the total moment of momentum of the Earth, $\boldsymbol{l}$, obeys the differential equation of III § 1.7:

$$\dot{\boldsymbol{l}} = \sum_1^n \boldsymbol{r}_i \times \boldsymbol{g}_i, \tag{1}$$

where

$$\boldsymbol{g}_i = Gmm_i \frac{\boldsymbol{r} - \boldsymbol{r}_i}{|\boldsymbol{r} - \boldsymbol{r}_i|^3}, \qquad i = 1, 2, \ldots, n; \tag{2}$$

and m denotes the mass and $\boldsymbol{r} = r\boldsymbol{u}$ the position vector of the Moon. (As in § 1, we take the origin of coordinates to be in the centre of mass of the Earth; the (x, y)-plane is the orbital plane of the Earth.)

It follows from (1) and (2) that

$$\begin{aligned}\dot{\boldsymbol{l}} &= Gm \sum_i m_i \frac{\boldsymbol{r}_i \times \boldsymbol{r}}{|\boldsymbol{r} - \boldsymbol{r}_i|^3} \\ &= \frac{Gm}{r^3} \sum_i m_i \boldsymbol{r}_i \times \boldsymbol{r}\left[1 + 3\frac{r_i}{r}\,\boldsymbol{u}_i\boldsymbol{u}\right],\end{aligned} \tag{3}$$

if we neglect terms of the second and higher orders in r_i/r in the square brackets. Now, since

$$\sum_i m_i \boldsymbol{r}_i = 0$$

and

$$\sum_i m_i r_i^2 (\boldsymbol{u}_i\boldsymbol{u})\boldsymbol{u}_i = \boldsymbol{J}\boldsymbol{u} \tag{4}$$

(compare the corresponding terms in equations (5) and (8) of § 1.1), equation (3) can be transformed into

$$\dot{\boldsymbol{l}} = \frac{3Gm}{r^3} (\boldsymbol{J}\boldsymbol{u}) \times \boldsymbol{u}. \tag{5}$$

We now write

$$\boldsymbol{l} = l \, . \, \boldsymbol{\lambda}, \tag{6}$$

where

$$\boldsymbol{\lambda} = \begin{pmatrix} -\sin \varepsilon \sin \psi \\ +\sin \varepsilon \cos \psi \\ \cos \varepsilon \end{pmatrix} \tag{7}$$

and $(90° + \psi)$ and $(90° - \varepsilon)$ are the polar coordinates of $\boldsymbol{\lambda}$. We define the vectors $\boldsymbol{\lambda}_\psi$ and $\boldsymbol{\lambda}_\varepsilon$ by

$$\left. \begin{aligned} \frac{\partial \boldsymbol{\lambda}}{\partial \psi} &\equiv \boldsymbol{\lambda}_\psi \sin \varepsilon, \\ \frac{\partial \boldsymbol{\lambda}}{\partial \varepsilon} &\equiv \boldsymbol{\lambda}_\varepsilon. \end{aligned} \right\} \tag{8}$$

Then

$$\dot{\boldsymbol{l}} = l\boldsymbol{\lambda}_\psi \sin \varepsilon \, . \, \dot{\psi} + l\boldsymbol{\lambda}_\varepsilon \, . \, \dot{\varepsilon} + \dot{l}\boldsymbol{\lambda}. \tag{9}$$

If we equate the expressions (5) and (9) and take the scalar products with $\boldsymbol{\lambda}_\psi$, $\boldsymbol{\lambda}_\varepsilon$, $\boldsymbol{\lambda}$, in turn, we obtain

$$\left. \begin{aligned} l \sin \varepsilon \, \dot{\psi} &= 3 \frac{Gm}{r^3} [\boldsymbol{Ju}, \boldsymbol{u}, \boldsymbol{\lambda}_\psi], \\ l \, . \, \dot{\varepsilon} &= 3 \frac{Gm}{r^3} [\boldsymbol{Ju}, \boldsymbol{u}, \boldsymbol{\lambda}_\varepsilon], \\ \dot{l} &= 3 \frac{Gm}{r^3} [\boldsymbol{Ju}, \boldsymbol{u}, \boldsymbol{\lambda}]. \end{aligned} \right\} \tag{10}$$

We now make the following three assumptions:

(i) The Earth is a rigid body.
(ii) It possesses rotational symmetry.
(iii) The direction $\boldsymbol{\lambda}$ of the moment of momentum $\boldsymbol{l}$ coincides with the axis of rotational symmetry.

Although these assumptions are not exactly correct, they are so to a high degree of approximation.

The volume products in equation (10) are independent of the orientation of the coordinate system to which we relate the three factors. To evaluate the products, therefore, we choose a coordinate system in which they become particularly simple.

Naturally we take the axis of rotational symmetry as the z-axis, so that (on the right-hand side) $\boldsymbol{\lambda} = \begin{pmatrix} 0 \\ 0 \\ 1 \end{pmatrix}$. We take the new x-axis in the original (x, y)-plane (in the orbital plane of the Earth). Now

$$\boldsymbol{\lambda}_\psi = -\begin{pmatrix} \cos\psi \\ \sin\psi \\ 0 \end{pmatrix}$$

lies in this plane and is always perpendicular to the new z-axis. In the new coordinates, therefore, $\boldsymbol{\lambda}_\psi$ is one of the vectors $\pm \begin{pmatrix} 1 \\ 0 \\ 0 \end{pmatrix}$. We choose the lower sign, so that in this coordinate system $\psi = 0$, $\varepsilon = 0$. From this it follows that

$$\boldsymbol{\lambda}_\varepsilon = \begin{pmatrix} 0 \\ 1 \\ 0 \end{pmatrix}.$$

In the new coordinates, the matrix $\boldsymbol{J}$ has the diagonal form § 1.1(10). It is obvious that only the new z-axis is fixed relatively to the body, but, on account of the rotational symmetry that has been assumed to exist about this axis, the quadratic form $\boldsymbol{u'Ju}$ must assume the form

$$J_x(u_x^2 + u_y^2) + J_z u_z^2.$$

With these stipulations equations (10) may be written

$$\left.\begin{aligned} l \sin\varepsilon\dot{\psi} &= 3\frac{Gm}{r^3}\begin{vmatrix} J_x u_x, & u_x, & 1 \\ J_y u_y, & u_y, & 0 \\ J_z u_z, & u_z, & 0 \end{vmatrix}, \\ l \qquad \dot{\varepsilon} &= 3\frac{Gm}{r^3}\begin{vmatrix} J_x u_x, & u_x, & 0 \\ J_y u_y, & u_y, & 1 \\ J_z u_z, & u_z, & 0 \end{vmatrix}, \\ \dot{l} \qquad &= 3\frac{Gm}{r^3}\begin{vmatrix} J_x u_x, & u_x, & 0 \\ J_y u_y, & u_y, & 0 \\ J_z u_z, & u_z, & 1 \end{vmatrix}. \end{aligned}\right\} \qquad (11)$$

The quantities on the left-hand side are referred to the original coordinate system fixed in space, while those on the right-hand side are referred to the new system whose z-axis is fixed with respect to the body. From (11) it follows that

$$\left.\begin{aligned} l \sin \varepsilon \dot{\psi} &= \frac{3Gm}{r^3}(J_y - J_z)u_y u_z, \\ l \quad \dot{\varepsilon} &= \frac{3Gm}{r^3}(J_z - J_x)u_z u_x, \\ \dot{l} \quad &= 0 \end{aligned}\right\} \tag{12}$$

the last on account of $J_x = J_y$, and, finally,

$$\left.\begin{aligned} l \sin \varepsilon \dot{\psi} &= +3\frac{Gm}{r^3}(C - A)u_y u_z, \\ l \quad . \dot{\varepsilon} &= -3\frac{Gm}{r^3}(C - A)u_x u_z \end{aligned}\right\} \tag{13}$$

by § 1.3(3).

2. The approximate theory just presented is very simple and yields an approximation which, as we shall see in § 2.5, is useful in practice. It suffers from a difficulty of principle, however, which considerably diminishes its value. The assumption 1(iii) was obviously made primarily on grounds of simplicity; but once it has been introduced, the theory can no longer be improved within its own framework—while formerly we have always preferred theories in which our first approximation can be improved by iteration to an arbitrary degree of accuracy. In what follows, therefore, we give up assumption (iii) and substitute it by an exact theory.

Our starting point is equation 1(5):

$$\dot{\boldsymbol{l}} = 3\frac{Gm}{r^3}(\boldsymbol{Ju}) \times \boldsymbol{u}. \tag{1}$$

In III § 1.7 $\boldsymbol{l}$ was defined by

$$\boldsymbol{l} = \sum_{i=1}^{n} \boldsymbol{r}_i \times m_i \dot{\boldsymbol{r}}_i. \tag{2}$$

We shall now transform this expression.

The vectors $\boldsymbol{r}_i$ rotate together with the planets. It can easily be shown geometrically that, in the time element $\mathrm{d}t$, $\boldsymbol{r}_i$ changes by

$$\mathrm{d}\boldsymbol{r}_i = \boldsymbol{\omega} \times \boldsymbol{r}_i \,.\, \mathrm{d}t,$$

where $\boldsymbol{\omega}$ denotes the vector of the instantaneous angular velocity. (The direction of $\boldsymbol{\omega}$ is that of the oriented instantaneous axis of rotation, and its magnitude is that of the instantaneous velocity of rotation.) It follows that

$$\dot{\boldsymbol{r}}_i = \boldsymbol{\omega} \times \boldsymbol{r}_i. \tag{3}$$

Now by (2)

$$\dot{\boldsymbol{l}} = \sum_i m_i \boldsymbol{r}_i \times \ddot{\boldsymbol{r}}_i, \tag{4}$$

so we must first evaluate $\ddot{\boldsymbol{r}}_i$. It may be obtained by differentiating (3):

$$\begin{aligned}\ddot{\boldsymbol{r}}_i &= \dot{\boldsymbol{\omega}} \times \boldsymbol{r}_i + \boldsymbol{\omega} \times \dot{\boldsymbol{r}}_i \\ &= \dot{\boldsymbol{\omega}} \times \boldsymbol{r}_i + \boldsymbol{\omega} \times (\boldsymbol{\omega} \times \boldsymbol{r}_i)\end{aligned}$$

(again by (3)); therefore, by means of the Lagrange identity III § 5.5(1),

$$\ddot{\boldsymbol{r}}_i = \dot{\boldsymbol{\omega}} \times \boldsymbol{r}_i - (\boldsymbol{\omega}^2)\boldsymbol{r}_i + (\boldsymbol{\omega}\boldsymbol{r}_i)\boldsymbol{\omega}. \tag{5}$$

If we set this expression in (4) and then apply the Lagrange identity once again, we obtain

$$\dot{\boldsymbol{l}} = \sum_i m_i\{\boldsymbol{r}_i^2\dot{\boldsymbol{\omega}} - (\boldsymbol{r}_i\dot{\boldsymbol{\omega}})\boldsymbol{r}_i + (\boldsymbol{\omega}\boldsymbol{r}_i)\boldsymbol{r}_i \times \boldsymbol{\omega}\}. \tag{6}$$

The reader may convince himself, by working out any one component of this vector equation explicitly, that

$$\dot{\boldsymbol{l}} = (3J\,U - \boldsymbol{J})\dot{\boldsymbol{\omega}} + (\boldsymbol{J}\boldsymbol{\omega}) \times \boldsymbol{\omega}, \tag{7}$$

where $\boldsymbol{J}$ and J have already been defined in § 1.1.

By comparing (1) and (7) we see that

$$(3J\,U - \boldsymbol{J})\dot{\boldsymbol{\omega}} + (\boldsymbol{J}\boldsymbol{\omega}) \times \boldsymbol{\omega} = 3\,\frac{Gm}{r^3}\,(\boldsymbol{J}\boldsymbol{u}) \times \boldsymbol{u}. \tag{8}$$

We shall now simplify this equation by bringing the matrix $\boldsymbol{J}$ into diagonal form, that is, as in § 1.1 and 2, we choose the principal axes of inertia of the rotating body as coordinate axes. At this point, however, a difficulty appears which must be removed. We cannot immediately refer equation (8) to the system of the principal axes of inertia, because this system (even in the case of rotational symmetry) can no longer be regarded as fixed in space. Equation (8), however, is only proved for systems fixed in space. To be able to derive the corresponding equation for systems fixed in the body, we must investigate the relative rotation of the two kinds of system more closely.

3. The rotation matrix $\boldsymbol{R}$, which gives the transformation between the coordinate system fixed in space and that defined by the principal axes of inertia, can be obtained similarly as in I § 4.3 and I § 5.6. We first rotate the system about the space fixed z-axis through the angle ψ, then through the angle ε about the new x-axis, and finally through the angle ϑ around the last obtained z-axis. (The angles ψ, ε, ϑ correspond to the angles Ω, ι, ω of I §§ 4.3 and 5.6.) These rotations may be represented by the matrices

$$\boldsymbol{\Psi} = \begin{pmatrix} \cos\psi & \sin\psi & 0 \\ -\sin\psi & \cos\psi & 0 \\ 0 & 0 & 1 \end{pmatrix}, \tag{1}$$

$$\mathbf{E} = \begin{pmatrix} 1 & 0 & 0 \\ 0 & \cos\varepsilon & \sin\varepsilon \\ 0 & -\sin\varepsilon & \cos\varepsilon \end{pmatrix}, \tag{2}$$

$$\boldsymbol{\Theta} = \begin{pmatrix} \cos\vartheta & \sin\vartheta & 0 \\ -\sin\vartheta & \cos\vartheta & 0 \\ 0 & 0 & 1 \end{pmatrix}, \tag{3}$$

and

$$\boldsymbol{R} = \boldsymbol{\Theta}\mathbf{E}\boldsymbol{\Psi}. \tag{4}$$

Any vector referred to the coordinate system fixed in the body we denote with the suffix B, then

$$\left.\begin{aligned} (\boldsymbol{r}_i)_B &= \boldsymbol{R}\boldsymbol{r}_i, \\ \boldsymbol{r}_i &= \boldsymbol{R}'(\boldsymbol{r}_i)_B, \end{aligned}\right\} \tag{5}$$

($\boldsymbol{R}' = \boldsymbol{R}^{-1}$ is the transpose matrix of $\boldsymbol{R}$),

$$\begin{aligned} \dot{\boldsymbol{r}}_i &= \dot{\boldsymbol{R}}'(\boldsymbol{r}_i)_B \\ &= \dot{\boldsymbol{R}}'\boldsymbol{R}\boldsymbol{r}_i. \end{aligned} \tag{6}$$

Now we write

$$\dot{\boldsymbol{R}}'\boldsymbol{R} \equiv \boldsymbol{\Omega} \tag{7}$$

so that

$$\dot{\boldsymbol{r}}_i = \boldsymbol{\Omega}\boldsymbol{r}_i. \tag{8}$$

Between the matrix $\boldsymbol{\Omega}$ and the vector of angular velocity, $\boldsymbol{\omega}$, there is a simple relation which we shall now derive.

First we must show that the matrix $\boldsymbol{\Omega}$ is skew-symmetric. By differentiating

$$\boldsymbol{R}'\boldsymbol{R} = \boldsymbol{U}$$

with respect to time, ($\boldsymbol{U}$ is the unit matrix), we obtain

$$\dot{\boldsymbol{R}}'\boldsymbol{R} + \boldsymbol{R}'\dot{\boldsymbol{R}} = 0,$$

that is

$$(\dot{\boldsymbol{R}}'\boldsymbol{R}) + (\dot{\boldsymbol{R}}'\boldsymbol{R})' = 0$$

or

$$\boldsymbol{\Omega} + \boldsymbol{\Omega}' = 0.$$

To derive the relation between $\boldsymbol{\Omega}$ and $\boldsymbol{\omega}$, we now write

$$\boldsymbol{\Omega} = \begin{pmatrix} 0 & -\Omega_z & \Omega_y \\ \Omega_z & 0 & -\Omega_x \\ -\Omega_y & \Omega_x & 0 \end{pmatrix}. \tag{9}$$

Then, as may be confirmed by simple algebra,

$$\boldsymbol{\Omega}\boldsymbol{r}_i = \begin{pmatrix} \Omega_x \\ \Omega_y \\ \Omega_z \end{pmatrix} \times \boldsymbol{r}_i. \tag{10}$$

By comparison with (8) it follows that

$$\dot{\boldsymbol{r}}_i = \begin{pmatrix} \Omega_x \\ \Omega_y \\ \Omega_z \end{pmatrix} \times \boldsymbol{r}_i. \tag{11}$$

On the other hand, by 2(3),

$$\dot{\boldsymbol{r}}_i = \boldsymbol{\omega} \times \boldsymbol{r}_i \tag{12}$$

and from the last two equations we obtain

$$\left.\begin{aligned} \omega_x &= \Omega_x, \\ \omega_y &= \Omega_y, \\ \omega_z &= \Omega_z, \end{aligned}\right\} \tag{13}$$

and

$$\boldsymbol{\Omega}\boldsymbol{r}_i = \boldsymbol{\omega} \times \boldsymbol{r}_i. \tag{14}$$

Equations (13) and (14) are the relations which we wished to derive.

The transformations (5) are valid not only for position vectors but for all vectors and so, in particular,

$$\begin{aligned}(\dot{\boldsymbol{r}}_i)_B &= \boldsymbol{R}\dot{\boldsymbol{r}}_i \\ &= \boldsymbol{R}\dot{\boldsymbol{R}}'(\boldsymbol{r}_i)_B \qquad (15)\end{aligned}$$

(by equation (5)). Completely analogous expressions to those for $\boldsymbol{\Omega}$ hold also for the matrix $\boldsymbol{R}\dot{\boldsymbol{R}}'$. We denote it by $\boldsymbol{\Omega}^*$. Now by (7),

$$\boldsymbol{\Omega}^* = \boldsymbol{R}\dot{\boldsymbol{R}}' = \boldsymbol{R}\boldsymbol{\Omega}\boldsymbol{R}'. \qquad (16)$$

In the same way as a unique vector $\boldsymbol{\omega}$ corresponds to the matrix $\boldsymbol{\Omega}$, (equation (13)), so a vector $\boldsymbol{\omega}^*$ corresponds to the matrix $\boldsymbol{\Omega}^*$. We assert

$$\boldsymbol{\omega}^* = (\boldsymbol{\omega})_B. \qquad (17)$$

In fact,

$$\begin{aligned}\boldsymbol{\omega}^* \times (\boldsymbol{r}_i)_B &= \boldsymbol{\Omega}^*(\boldsymbol{r}_i)_B \\ &= (\boldsymbol{R}\boldsymbol{\Omega}\boldsymbol{R}')(\boldsymbol{R}\boldsymbol{r}_i) \quad \text{(by (16) and (5))} \\ &= \boldsymbol{R}(\boldsymbol{\Omega}\boldsymbol{r}_i) \\ &= \boldsymbol{R}(\boldsymbol{\omega} \times \boldsymbol{r}_i) \\ &= (\boldsymbol{R}\boldsymbol{\omega}) \times (\boldsymbol{R}\boldsymbol{r}_i),\end{aligned}$$

(the vector product of $\boldsymbol{\omega}$ and $\boldsymbol{r}_i$ undergoes the same rotation $\boldsymbol{R}$ as do its factors) so that, finally,

$$\boldsymbol{\omega}^* \times (\boldsymbol{r}_i)_B = (\boldsymbol{\omega})_B \times (\boldsymbol{r}_i)_B \qquad (18)$$

for all vectors $(\boldsymbol{r}_i)_B$. From this follows the assertion (17).

Expressed in words this means that the vector $\boldsymbol{\omega}$, which belongs to the matrix $\boldsymbol{\Omega} = \dot{\boldsymbol{R}}'\boldsymbol{R}$ is the angular velocity of the body in the coordinate system fixed in space. The vector $\boldsymbol{\omega}^*$, which belongs to the matrix $\boldsymbol{\Omega}^* = \boldsymbol{R}\dot{\boldsymbol{R}}' = \boldsymbol{R}\boldsymbol{\Omega}\boldsymbol{R}'$, is the vector $\boldsymbol{\omega}_B = \boldsymbol{R}\boldsymbol{\omega}$, that is the vector of the instantaneous angular velocity referred to our system of coordinates fixed with respect to the body.

4. We are now in a position to write the dynamical equation 2(8),

$$(3JU - \boldsymbol{J})\dot{\boldsymbol{\omega}} + (\boldsymbol{J}\boldsymbol{\omega}) \times \boldsymbol{\omega} = 3\,\frac{Gm}{r^3}\,(\boldsymbol{J}\boldsymbol{u}) \times \boldsymbol{u}, \qquad (1)$$

in the system of coordinates fixed in the body. First, it follows from

$$\boldsymbol{u}'\boldsymbol{J}\boldsymbol{u} = \boldsymbol{u}'_B\boldsymbol{J}_B\boldsymbol{u}_B \tag{2}$$

and

$$\boldsymbol{u}_B = \boldsymbol{R}\boldsymbol{u}, \tag{3}$$

that

$$\boldsymbol{J} = \boldsymbol{R}'\boldsymbol{J}_B\boldsymbol{R}. \tag{4}$$

The scalar

$$J = \tfrac{1}{3}\sum_i m_i \boldsymbol{r}_i^2, \tag{5}$$

on the other hand, is invariant with respect to the rotation, so that (1) may be written:

$$\boldsymbol{R}'(3J\boldsymbol{U} - \boldsymbol{J}_B)\boldsymbol{R}\dot{\boldsymbol{\omega}} + (\boldsymbol{R}'\boldsymbol{J}_B\boldsymbol{R}\boldsymbol{\omega}) \times \boldsymbol{\omega}$$
$$= 3\,\frac{Gm}{r^3}\,(\boldsymbol{R}'\boldsymbol{J}_B\boldsymbol{R}\boldsymbol{u}) \times \boldsymbol{u} \tag{6}$$

or, premultiplying by $\boldsymbol{R}$,

$$(3J\boldsymbol{U} - \boldsymbol{J}_B)\boldsymbol{R}\boldsymbol{\omega} + (\boldsymbol{J}_B \cdot \boldsymbol{R}\boldsymbol{\omega}) \times (\boldsymbol{R}\boldsymbol{\omega})$$
$$= 3\,\frac{Gm}{r^3}\,(\boldsymbol{J}_B \cdot \boldsymbol{R}\boldsymbol{u}) \times (\boldsymbol{R}\boldsymbol{u}). \tag{7}$$

Now $\boldsymbol{R}\boldsymbol{\omega} = \boldsymbol{\omega}_B$, and, further,

$$\begin{aligned}\boldsymbol{R}\dot{\boldsymbol{\omega}} &= \boldsymbol{R}(\boldsymbol{R}'\boldsymbol{\omega}_B)^{\cdot} = \boldsymbol{R}\dot{\boldsymbol{R}}'\boldsymbol{\omega}_B + \boldsymbol{R}\boldsymbol{R}'\dot{\boldsymbol{\omega}}_B \\ &= \boldsymbol{\Omega}_B\boldsymbol{\omega}_B + \dot{\boldsymbol{\omega}}_B \qquad \text{(by 3(16) and 3(17))} \\ &= \boldsymbol{\omega}_B \times \boldsymbol{\omega}_B + \dot{\boldsymbol{\omega}}_B \\ &= \dot{\boldsymbol{\omega}}_B.\end{aligned} \tag{9}$$

By means of equations (3), (8) and (9), therefore, we can transform equation (7) into

$$(3J\boldsymbol{U} - \boldsymbol{J}_B)\dot{\boldsymbol{\omega}}_B + (\boldsymbol{J}_B\boldsymbol{\omega}_B) \times \boldsymbol{\omega}_B$$
$$= 3\,\frac{Gm}{r^3}\,(\boldsymbol{J}_B\boldsymbol{u}_B) \times \boldsymbol{u}_B. \tag{10}$$

This equation has exactly the same form as equation (1) has in the coordinate system fixed in space. If we write out its three components and introduce the principal moments of inertia of § 1.3(3)

$$A = \sum_i m_i(y_i^2 + z_i^2), \text{ etc.}, \tag{11}$$

we obtain Euler's *dynamical equations*

$$\left.\begin{aligned}A\dot{\omega}_x + (C - B)\omega_y\omega_z &= 3\,\frac{Gm}{r^3}\,(C - B)u_y u_z,\\ B\dot{\omega}_y + (A - C)\omega_z\omega_x &= 3\,\frac{Gm}{r^3}\,(A - C)u_z u_x,\\ C\dot{\omega}_z + (B - A)\omega_x\omega_y &= 3\,\frac{Gm}{r^3}\,(B - A)u_x u_y\end{aligned}\right\} \tag{12}$$

for the rotation of a rigid body; here ω_x, ω_y and ω_z are the components of ω_B. In this derivation rotational symmetry has not been assumed. The particular form 1(2) has been assumed, however, for the external forces $\boldsymbol{g}_i$ acting on the body. For general forces $\boldsymbol{g}_i$, the components of the vector $\sum_i \boldsymbol{r}_i \times \boldsymbol{g}_i$, referred to the coordinate system fixed in the body, must be substituted for the right-hand sides of equations (12).

5. In Euler's three dynamical equations 4(12), there are six unknowns: ω_x, ω_y, ω_z, and the three angles ψ, ε, ϑ (which are contained in the components of $\boldsymbol{u}_B = \boldsymbol{R}\boldsymbol{u}$). For these we must have three further equations. We already have them, implicitly, in equation 3(16) which, with the help of 3(17), we may write in the form

$$\boldsymbol{\Omega}_B = \boldsymbol{R}\dot{\boldsymbol{R}}'. \tag{1}$$

By 3(4)

$$\boldsymbol{R} = \boldsymbol{\Theta}\mathbf{E}\boldsymbol{\Psi}, \tag{2}$$

setting this in equation (1) we obtain

$$\boldsymbol{\Omega}_B = \boldsymbol{\Theta}\mathbf{E}\boldsymbol{\Psi}(\boldsymbol{\Psi}'\mathbf{E}'\boldsymbol{\Theta}')^{\cdot} \tag{3}$$

or

$$\boldsymbol{\Omega}_B = (\boldsymbol{\Theta}\mathbf{E})\left(\boldsymbol{\Psi}\,\frac{\partial\boldsymbol{\Psi}'}{\partial\psi}\right)(\boldsymbol{\Theta}\mathbf{E})'\,.\,\dot{\psi} + \boldsymbol{\Theta}\left(\mathbf{E}\,\frac{\partial\mathbf{E}'}{\partial\varepsilon}\right)\boldsymbol{\Theta}'\,.\,\dot{\varepsilon} + \left(\boldsymbol{\Theta}\,\frac{\partial\boldsymbol{\Theta}'}{\partial\vartheta}\right).\,\dot{\vartheta}. \tag{4}$$

Written out in its single elements this equation reads:

$$\left.\begin{aligned}\omega_x &= \dot{\psi}\sin\vartheta\sin\varepsilon + \dot{\varepsilon}\cos\vartheta,\\ \omega_y &= \dot{\psi}\cos\vartheta\sin\varepsilon - \dot{\varepsilon}\sin\vartheta,\\ \omega_z &= \dot{\psi}\qquad\cos\varepsilon\qquad\qquad + \dot{\vartheta}.\end{aligned}\right\} \tag{5}$$

We have made use of the representations 3(1) – (3) for the matrices $\mathbf{\Psi}$, $\mathbf{E}$, $\mathbf{\Theta}$. From the first two equations we see immediately that

$$\left.\begin{aligned} \dot{\psi} \sin \varepsilon &= \omega_x \sin \vartheta + \omega_y \cos \vartheta, \\ \dot{\varepsilon} &= \omega_x \cos \vartheta - \omega_y \sin \vartheta, \end{aligned}\right\} \tag{6}$$

while, from the third equation,

$$\dot{\vartheta} = \omega_x \sin \vartheta \cot \varepsilon - \omega_y \cos \vartheta \cot \varepsilon + \omega_z. \tag{7}$$

If the geometrical relations (6) and (7) are taken together with the dynamical relations 4(12), we have the required system of six differential equations for the six unknown functions ω_x, ω_y, ω_z, ϑ, ε, ψ.

6. We can now proceed to the integration of the complete system of the Euler equations. They now read, by 4(12), 5(6) and (7),

$$\left.\begin{aligned} A\dot{\omega}_x + (C - B)\omega_y\omega_z &= 3\,\frac{Gm}{r^3}\,(C - B)u_y u_z, \\ B\dot{\omega}_y + (A - C)\omega_z\omega_x &= 3\,\frac{Gm}{r^3}\,(A - C)u_z u_x, \\ C\dot{\omega}_z + (B - A)\omega_x\omega_y &= 3\,\frac{Gm}{r^3}\,(B - A)u_x u_y, \\ \dot{\psi} \sin \varepsilon &= \omega_x \sin \vartheta + \omega_y \cos \vartheta, \\ \dot{\varepsilon} &= \omega_x \cos \vartheta - \omega_y \sin \vartheta, \\ \dot{\vartheta} &= (\omega_x \sin \vartheta - \omega_y \cos \vartheta) \cot \varepsilon + \omega_z. \end{aligned}\right\} \tag{1}$$

In these equations u_x, u_y, u_z are components (referred to a coordinate system fixed in the Earth) of the unit vector in the direction of the Moon:

$$(\boldsymbol{u})_B = \boldsymbol{R}\boldsymbol{u} \tag{2}$$

and $\boldsymbol{u}$ is given by lunar theory, so that the right-hand sides of the first three equations depend on t, ϑ, ε, ψ, but not on ω_x, ω_y, ω_z.

Again, as in § 1, we assume the Earth to be a rotationally symmetric rigid body, so that $A = B$. Then, from the third equation (1), it follows that

$$\omega_z = \text{const.} \tag{3}$$

The first two equations are linear in ω_x and ω_y. Suppose now that suitable approximations for ϑ, ε, ψ are known; (e.g. constants or linear expressions in time). Then these two equations can be separated from the rest and solved by the method for inhomogeneous linear equations given in III § 3.6. With this method we must first solve the linear homogeneous system

$$\left.\begin{aligned} \dot{\omega}_x + \omega_z \frac{C-A}{A} \cdot \omega_y &= 0, \\ \dot{\omega}_y - \omega_z \frac{C-A}{A} \cdot \omega_x &= 0. \end{aligned}\right\} \tag{4}$$

The method of III § 3.7 gives

$$\left.\begin{aligned} \omega_x &= \alpha\omega_z \cos(\nu t + \delta), \\ \omega_y &= \alpha\omega_z \sin(\nu t + \delta), \end{aligned}\right\} \tag{5}$$

for the solution, where α and δ are constants of integration and

$$\nu = \frac{C-A}{A}\,\omega_z. \tag{6}$$

The direction of the vector $(\omega_x, \omega_y, \omega_z)$ is the direction of the instantaneous axis of rotation. This therefore describes a circular cone about the axis of symmetry of the Earth. The semi-vertical angle of the cone is α, and the period of the motion

$$\begin{aligned} \frac{2\pi}{\nu} &= \frac{2\pi}{\omega_z}\frac{A}{C-A}\,[\text{years}] \\ &= \frac{A}{C-A}\,[\text{sidereal days}]. \end{aligned} \tag{7}$$

The possibility of this "free motion" of the Earth's axis (of the "pole of rotation") is quite independent of the existence of any disturbing force (or moment) acting on the axis of the Earth, whether produced by the Sun or the Moon. Whether it actually appears depends on the initial conditions which, in this case, are cosmogonical conditions. We shall briefly survey the observational results in § 9.

7. The two first equations of 6(1):

$$\left.\begin{aligned} \dot{\omega}_x + \omega_z \frac{C-A}{A} \cdot \omega_y &= 3 \frac{Gm}{r^3} \frac{C-A}{A} u_z u_y, \\ \dot{\omega}_y - \omega_z \frac{C-A}{A} \cdot \omega_x &= -3 \frac{Gm}{r^3} \frac{C-A}{A} u_x u_z, \end{aligned}\right\} \tag{1}$$

may be solved by the method of variation of parameters of III § 3.6, if their right-hand sides are regarded as known functions of the time. The exact working is long (but without any difficulties of principle) so we shall satisfy ourselves, here, with an approximate solution. We introduce into equations (1) the expressions for ω_x and ω_y in 5(6) and we neglect all the terms which do not contain the factor $\omega_z \approx \dot{\vartheta}$. This seems permissible since, without doubt, all other velocities are very small compared with

$$\omega_z \approx 2\pi \bigg/ \frac{1}{366{\cdot}25}. \tag{2}$$

With this approximation we obtain

$$\left.\begin{aligned} \dot{\omega}_x &= \dot{\psi} \cos \vartheta \,.\, \omega_z \sin \varepsilon - \dot{\varepsilon} \sin \vartheta \,.\, \omega_z = \omega_z \omega_y, \\ \dot{\omega}_y &= -\dot{\psi} \sin \vartheta \,.\, \omega_z \sin \varepsilon - \dot{\varepsilon} \cos \vartheta \,.\, \omega_z = -\omega_z \omega_x, \end{aligned}\right\} \tag{3}$$

and equations (1) transform into

$$\left.\begin{aligned} \dot{\psi} \sin \varepsilon \,.\, \cos \vartheta - \dot{\varepsilon} \sin \vartheta &= 3 \frac{Gm}{r^3} \frac{C-A}{C} \frac{u_z}{\omega_z} u_y, \\ \dot{\psi} \sin \varepsilon \,.\, \sin \vartheta + \dot{\varepsilon} \cos \vartheta &= 3 \frac{Gm}{r^3} \frac{C-A}{C} \frac{u_z}{\omega_z} u_x. \end{aligned}\right\} \tag{4}$$

If we use the expression 3(1) for $\mathbf{\Theta}$, we can write this system of equations in the form

$$\mathbf{\Theta} \begin{pmatrix} \dot{\varepsilon} \\ \dot{\psi} \sin \varepsilon \\ 0 \end{pmatrix} = 3 \frac{Gm}{r^3} \frac{C-A}{C} \frac{(u_z)_B}{\omega_z} \begin{pmatrix} u_x \\ u_y \\ 0 \end{pmatrix}_B, \tag{5}$$

where the suffix B again indicates reference to the system of coordinates fixed in the body. Since, by 4(2),

$$(\boldsymbol{u})_B = \boldsymbol{R}\boldsymbol{u} = \mathbf{\Theta}\mathbf{E}\mathbf{\Psi}\boldsymbol{u}, \tag{6}$$

it follows from (5) that

$$\begin{pmatrix} \dot{\varepsilon} \\ \dot{\psi} \sin \varepsilon \\ 0 \end{pmatrix} = 3 \frac{Gm}{r^3} \frac{C-A}{A} \frac{(u_z)_B}{\omega_z} \mathbf{E}\mathbf{\Psi} \begin{pmatrix} u_x \\ u_y \\ 0 \end{pmatrix}. \tag{7}$$

This, essentially, is equation 1(13). Only the notation is changed a little. The quantity $(u_z)_B$ is independent of ϑ and therefore equal to the third component of $\mathbf{E\Psi}\boldsymbol{u}$; the vector $\mathbf{E\Psi}\boldsymbol{u} = \boldsymbol{\Theta}^{-1}(\boldsymbol{u})_B$ is identical with the vector $\boldsymbol{u}$ in formulae 1(11)–(13). We can now apply formula 1(13) without any further hesitation. If its accuracy does not prove sufficient, we know how to improve it.

8. We shall now use equations 7(7), which we have just derived, to consider the influence of the Moon. For the sake of simplicity, we shall restrict ourselves to the representation of the orbit of the Moon as a circle, with radius a and a very small inclination ι. We may then place the perigee direction in the nodal line and we obtain

$$\boldsymbol{u} = \begin{pmatrix} \cos(\Omega + M) \\ \sin(\Omega + M) \\ \iota \sin M \end{pmatrix}. \tag{1}$$

Here Ω is again the longitude of the node. We also substitute zero order approximations for ψ, ε, ϑ on the right-hand side of equation 5(7); they are as follows:

$$\left.\begin{aligned} \psi &= 0, \\ \varepsilon &= \text{const.}, \\ \vartheta &= \vartheta_0 + \omega_z \,.\, t, \qquad \vartheta_0 = \text{const.} \end{aligned}\right\} \tag{2}$$

Since $\psi = 0$, we have $\mathbf{\Psi} = \boldsymbol{U}$, so that

$$\mathbf{E\Psi}\boldsymbol{u} = \mathbf{E}\boldsymbol{u}$$

$$= \begin{pmatrix} \cos(\Omega + M) \\ \cos\varepsilon \sin(\Omega + M) \\ -\sin\varepsilon \sin(\Omega + M) \end{pmatrix} + \iota \begin{pmatrix} 0 \\ \sin\varepsilon \\ \cos\varepsilon \end{pmatrix} \sin M. \tag{3}$$

Then we write

$$\left.\begin{aligned} M &= M_0 + n(t - t_0), \\ \Omega &= \Omega_0 + \dot{\Omega}_0(t - t_0), \end{aligned}\right\} \tag{4}$$

(the latter equation holds approximately by III § 6.4). Now on the right-hand side of equation 7(7) there are terms which are proportional to

$$\begin{array}{lll} 1, & \cos\Omega, & \sin\Omega, \\ & \cos(\Omega + 2M), & \sin(\Omega + 2M), \\ & \cos(2\Omega + 2M), & \sin(2\Omega + 2M) \end{array}$$

and whose factors of proportionality are all of the same order. Integration with respect to time, however, destroys this equality in orders of magnitude: each individual term is multiplied by a factor proportional to its period. (Compare the analogous small divisors of planetary theory in III § 2.3.) Now $|\dot{\Omega}_0| \ll n$, so we shall be satisfied here with the terms given by the integration of 1, cos Ω and sin Ω alone. Thus we obtain

$$\left.\begin{aligned} \varepsilon(t) &= \varepsilon_0 - \frac{3}{2}\frac{Gm}{a^3}\frac{C-A}{C}\frac{\cos\varepsilon}{\omega_z}\frac{\iota}{\dot{\Omega}_0}\cos\Omega, \\ \psi(t) &= -\frac{3}{2}\frac{Gm}{a^3}\frac{C-A}{C}\frac{\cos\varepsilon}{\omega_z}\left[(t-t_0)+2\cot 2\varepsilon \times \frac{\iota}{\dot{\Omega}_0}\sin\Omega\right]. \end{aligned}\right\} \qquad (5)$$

The secular term is the precession and the periodic terms are the nutation of the axis of the Earth. In the conventional coordinate system $\psi(t)$ has the opposite sign to that which it has in our space-fixed coordinate system. (See I § 1.2.) The conventional coordinate system (the second equatorial system) is related with our space-fixed system by the matrix transformation $\mathbf{E\Psi}$ (while the first equatorial system is related with it by the transformation $\mathbf{\Theta E \Psi}$).

Equation (5) gives only the very first approximation. We recall once more the assumptions which we made to shorten the working. In Euler's equations 4(12) we have neglected all terms not containing ω_z. The orbit of the Moon has been assumed to be circular and its inclination to be so small that its square is negligible. Only the secular terms on the right-hand sides of equations 7(7) were considered. The first step in the sequence of successive approximations has been assumed to produce a useful approximation.

In practice this approximation is not sufficient, although all the neglected terms are, together, very small. Apart from the labour of the computation, however, there are no difficulties which have not been covered by the theory.

9. The observational value of the precession produced by the Sun and Moon together is about $-50''{\cdot}37(t-t_0)$. By 8(5) we have

$$\text{Precession} = -\frac{3}{2}\frac{Gm}{a^3}\frac{C-A}{A}\frac{\cos\varepsilon}{\omega_z}(t-t_0). \qquad (1)$$

The components due to the Moon and Sun are, therefore, in the ratio

$$\frac{m_☽}{a_☽^3} : \frac{m_\odot}{a_\odot^3} = \frac{m_☽}{a_☽^3} \approx 2{\cdot}16$$

from the data of III § 6.6:

$$a_☽ = 2{\cdot}57 \times 10^{-3},$$

$$m_☽ = \frac{1}{80{\cdot}3}\,(\text{mass of Earth} + \text{Moon})$$

$$= \frac{1}{80{\cdot}3} \times 3{\cdot}0 \times 10^{-6}.$$

The influence of the Moon, therefore, is about twice as great as that of the Sun. The total precession is

$$-\tfrac{3}{2}\,G\left(1 + \frac{m_☽}{a_☽^3}\right)\frac{C - A}{C}\,\frac{\cos\varepsilon}{\omega_z}\,(t - t_0).$$

Comparison with the empirical value yields

$$\frac{C - A}{C} = \frac{1}{305{\cdot}1}\,. \tag{2}$$

The variation in the "obliquity of the ecliptic" (the inclination of the Earth's orbital and equatorial planes) is, by 8(5),

$$\varepsilon - \varepsilon_0 = \left(-\frac{3}{2}\,\frac{Gm_☽}{a_☽^3}\,\frac{C - A}{C}\,\frac{\cos\varepsilon}{\omega_z}\right)\frac{\iota_☽\cos\Omega}{\dot{\Omega}_0}$$

$$= \left(-50''{\cdot}37 \times \frac{2{\cdot}16}{1 + 2{\cdot}16}\right) \times \frac{\iota_☽\cos\Omega}{\dot{\Omega}_0}$$

$$= 9''{\cdot}21\cos\Omega. \tag{3}$$

(The solar component vanishes since $\iota_\odot = 0$.) The periodic variation in ψ is also given by 8(5) and is $-17''\sin\Omega$. In practice, the coefficient of $\cos\Omega$ in (3) is determined directly by empirical means. Between the values obtained from precession observations and from the mass of the Moon, there is a small but significant difference, the origin of which does not yet appear to have been fully explained.

From the value of $(C - A)/C$, equation (2), the period 6(7) of the free motion of the Earth's axis may be evaluated. Its (approximate) period in sidereal days is

$$\frac{A}{C - A} = \frac{C}{C - A} - 1, \tag{4}$$

that is about 304 sidereal or 303 mean solar days.

This motion of the axis of rotation relative to the axis of symmetry of the Earth has not yet proved observable. Instead, a periodic oscillation of the axis of rotation, with an amplitude of the order of $1/10''$ has been found, with two periods of roughly twelve and fourteen months superimposed on each other. The annual period is explained by yearly mass displacements on the surface of the Earth; such as might be caused, for example, by the Monsoon. The fourteen month period is thought to be a distortion of the ten month period (4). It is supposed that the elasticity of the Earth may be the cause, but the matter does not yet seem to be completely clear.

3. The rotation of the Moon

1. To investigate the rotation of the Moon and the influence of the Earth on it, we must exchange the roles of Earth and Moon in § 2. Euler's dynamical and geometrical equations (§ 2.4(1)) will still hold. In this case, however, we cannot put $A = B$. It is known from observations that, except for small periodic oscillations, the Moon always turns the same face towards us. These small oscillations are partly geometrical. For example, the Moon rotates about its axis with a constant angular velocity; the Earth moves around the Moon with exactly the same mean angular velocity; its instantaneous angular velocity, however, deviates from this mean because of the eccentricity of the Earth's orbit around the Moon. As a result sometimes one, sometimes the other limb of the Moon is turned towards the observer on the Earth, who sees a small pendular oscillation about the axis with a period of one month. In addition to this, however, there are small variations in the angular velocity of the Moon, which are not to be explained geometrically, but dynamically, as small deviations from its mean rotation From this we conclude that the mass distribution in the Moon is somewhat elongated in the direction of the Earth, so that

the Moon always tends to point in this preferred direction along which an external force is acting, just like a magnetic needle. (This is in agreement with the following idea: the gravitational action of the Earth raised a tidal wave on the Moon, while it was still molten. Both crests of this wave always lay on the line joining the Earth and Moon: one pointed towards the Earth, and the other away from it. Tidal friction forced the period of rotation of the Moon to coincide with that of its revolution around the Earth.) The observed position of stability suggests that the mean direction of the Earth should be the direction of a principal axis of inertia, say the x-direction of our coordinate system fixed in the body. Then the distances of the particles of the Moon from the x-axis are, on the average, smallest. By § 1.3(3), therefore, we expect

$$A < B, \qquad A < C. \tag{1}$$

2. The rotation of the Moon may be treated analytically in the following simple manner. The observations suggest that the actual rotation, about the z-axis fixed in the Moon, is governed by the equations

$$\left.\begin{aligned} \mathring{\omega}_x &= 0, \quad \mathring{\omega}_y = 0, \quad \mathring{\omega}_z = 2\pi/\text{sidereal month} \\ \mathring{\psi} &= 0, \quad \mathring{\varepsilon} = 0, \quad \mathring{\vartheta} = \mathring{\omega}_z t \end{aligned}\right\} \tag{1}$$

(By the z-axis we mean the principal axis of inertia which is roughly perpendicular to the orbital plane of the Earth.) We now set up approximate equations for the actual rotation by the method of III § 3.8, while we linearize Euler's equations § 2.4(12) "around the approximate solution (1)." We then readily obtain for $\delta\omega_x = \omega_x - \mathring{\omega}_x$, etc.,

$$\left.\begin{aligned} (\delta\omega_x)^{\cdot} + \frac{C-B}{A}\,\mathring{\omega}_z\,.\,\delta\omega_y &= 3\,\frac{Gm}{r^3}\,\frac{C-B}{A}\,u_y u_z \\ (\delta\omega_y)^{\cdot} + \frac{A-C}{B}\,\mathring{\omega}_z\,.\,\delta\omega_x &= 3\,\frac{Gm}{r^3}\,\frac{A-C}{B}\,u_z u_x \\ (\delta\omega_z)^{\cdot} \qquad\qquad &= 3\,\frac{Gm}{r^3}\,\frac{B-A}{C}\,u_x u_y \end{aligned}\right\} \tag{2}$$

and three equations which arise from the linearization of the geometrical equations. The expressions $(u_y u_z)$ and so on, are obtained from the equation

$$(\boldsymbol{u})_B = \boldsymbol{R}(\boldsymbol{u})_{\text{space-fixed}}$$
$$= \left(\boldsymbol{R} + \frac{\partial \boldsymbol{R}}{\partial \vartheta}\,\delta\vartheta + \frac{\partial \boldsymbol{R}}{\partial \varepsilon}\,\delta\varepsilon + \frac{\partial \boldsymbol{R}}{\partial \psi}\,\delta\psi\right)_{\begin{pmatrix}\vartheta=\mathring{\vartheta}\\ \varepsilon=0\\ \psi=0\end{pmatrix}} \boldsymbol{u}.$$

There are, altogether, six linear differential equations with constant or periodic coefficients for the six unknown functions $\delta\omega_x$, $\delta\omega_y$, $\delta\omega_z$, $\delta\vartheta$, $\delta\varepsilon$, $\delta\psi$. This system can be solved by the methods of III § 3. Then, by comparison with observation, the quantities $\dfrac{C - B}{A}$ etc. may be evaluated.

We shall not, here, go any further into this method; instead we shall be satisfied with a qualitative inference from equations (2). Suppose we have solved the complete system of six equations. Then the right-hand sides of equations (2) can be regarded as known functions of the time; (the left-hand sides are such functions also, but that is of no importance at the moment). The first two equations then form an inhomogeneous linear system for the functions $\delta\omega_x$, $\delta\omega_y$. By III § 3.6 the general solution of the inhomogeneous system is the sum of any one of its particular solutions and the general solution of the homogeneous system. If the solution of the inhomogeneous system is to be stable, (III § 3.8), the solution of the homogeneous system must be also. This can only be so, if all the roots of the characteristic equation III § 3.7(2) are imaginary. This equation reads, in this case,

$$\begin{vmatrix} \lambda, & \dfrac{C - B}{A} \\ \dfrac{A - C}{B}, & \lambda \end{vmatrix} = 0$$

or

$$\lambda^2 = \frac{C - B}{A}\,\frac{A - C}{B}.$$

From the requirement $\lambda^2 < 0$, it follows that, either,

$$C > A, \qquad C > B,$$

or,

$$C < A, \qquad C < B.$$

The possibility $A > C$ is excluded by 1(1), therefore,

$$A < B < C, \tag{3}$$

the Moon is flattened along its axis of rotation, as would be expected in view of the centrifugal forces acting on it.

4. Summary and conclusion

In this chapter the celestial bodies were considered as systems of mass points whose distances from each other remain constant for all time. In § 1 it was shown that differences between the principal moments of inertia of such a body give rise to a disturbing force which will, for example, cause a periastron motion of a satellite orbit.

In § 2 we investigated the influence of the Moon and Sun on the rotation of the Earth. At first we supposed the direction of rotational momentum to coincide with the direction of the axis of (rotational) symmetry. An exact theory is given by Euler's dynamical and geometrical equations for the rotation of a rigid body. We apply this theory for the Sun and Moon and, under a few simplifying assumptions, we obtain first approximations for precession and nutation, retaining amongst the periodic terms only those with very small divisors. The approximation is satisfactory when compared with the rigorous theory. Between the rigorous theory and the observations, however, there is a small discrepancy which has not yet been explained. There is considerable difference between theory and observations of the free motion of the Earth's axis. This is usually explained by the elasticity of the Earth.

Finally, in § 3, we derived inequalities for the principal moments of inertia of the Moon from qualitative considerations of stability. This treatment is only a consideration of plausibility. For an exact quantitative treatment, we suggest the linearization of Euler's equations.

Further problems of celestial mechanics (the tides, the shape of the celestial bodies, etc.) need to be more closely related to physical considerations. They lie, therefore outside the scope of this book.

Exercises

1. The Sun possesses an axial rotation, whose direction coincides with that of the revolution of the planets and whose period is about 25 days. Let

the axis of rotation be the z-axis of a system of coordinates fixed in the body, then $C > A$. Can the motion of the perihelion of Mercury (III § 7) be explained by the theory of § 1.3? (Calculate $C - A$, and give an upper limit for A and, therefore, a lower limit for $(C - A)/A$. The radius of the Sun is 7×10^5 km. It shows no appreciable ellipticity.)

2. Clarify, by qualitative arguments, the reason for the sign of the right-hand side of the formula § 1.3(9).

3. Let the orbit of a satellite be perpendicular to the equatorial plane of a planet. What displacement will its periastron suffer (§ 1)?

NOTES

§ **2.1** and **6.** A very elementary treatment is given by

[26] Schürer, M., Die Präzession und Nutation der Erde. *El. d. Math.*, **3**, 54–59, 1948.

§ **2.4.** The expression "sidereal day" in formula (7) is to be understood quite literally; that is, it refers to the rotation of the Earth relatively to the celestial sphere, not, as in practical astronomy, relatively to the vernal equinox (the oriented line of intersection of the orbital and equatorial planes of the Earth).

§ **2.7.** The data were not given with great accuracy, so there was no need to state the epoch.

On the influence of the elasticity of the Earth on its rotation, see

[27] Jeffreys, H., *The Earth*, Third edition, Cambridge, 1952.

§ **3.2.** On the free rotation of the Moon see

[28] Jeffreys, H., *The Moon's Libration in Longitude*, *M.N.*, **117**, 475, 1957.

INDEX